YOGA & MEDITATION

A HOLISTIC APPROACH
TO PERFECT HOMEOSTASIS

YOGA
&
MEDITATION

A HOLISTIC APPROACH
TO PERFECT HOMEOSTASIS

DR YOGINI SHUBH VEER

PH.D. (Y, M & A.M.) USA

PH.D. (A.M.) INDIA

D.SC. (A.M.) INDIA

Matador
9 Priory Business Park,
Wistow Road, Kibworth Beauchamp,
Leicestershire. LE8 0RX
Tel: (+44) 116 279 2299
Fax: (+44) 116 279 2277
Email: books@troubador.co.uk
Web: www.troubador.co.uk/matador

ISBN 978 1780883 069

British Library Cataloguing in Publication Data.
A catalogue record for this book is available from the British Library.

Typeset in 11pt Bembo by Troubador Publishing Ltd, Leicester, UK

Matador is an imprint of Troubador Publishing Ltd

DEDICATED
WITH ALL LOVE AND GRATITUDE
To
MY GURUS
Shri SURYANARAIN
&
Shri MAHAVEER

All yoga postures shown in the book are by the author, when aged above 40.

Photographs were taken by the author's students.

CONTENTS

PREFACE

Achieving and maintaining homeostasis is the prime objective of every living being. The holistic approach through Yoga and Meditation appears to be the one method which conditions the system, leading to more enduring results, for recovering, maintaining and enhancing this most vital state of homeostasis.

Devised some four decades ago, this yoga and meditation course with its holistic approach has been taught to thousands with excellent results. It has served its purpose and it still continues to attract new adepts. Whatever be the need or creed, there is something to suit one and all.

The course was initially offered as a series of lessons in hatha yoga and Siddha kundalini mahayoga meditation. It was meant for those interested in the metaphysical and spiritual form of yoga. The aim was to bring within reach of a maximum of seekers, this ancient path of holistic living.

Gradually, it turned out that many of those attending the courses had personal problems, either health, family or professional. They were more in search of some relief than wanting to become adepts in yoga and meditation. Hundreds among came for stress related conditions, the curse of modern living. We had to adapt to this reality.

The remarkable results obtained through yoga and meditation are quite impressive, simply astounding. The majority has been cured or got significant relief. Among are cases where traditional medicine had failed. Some extracts from personal testimonies are mentioned.

Repeated requests to have the course published have given shape to this present work. The aim is to allow greater access to this age-old wisdom.

May it serve its holistic purpose!

Dr Yogini Shubh Veer
27th September 2012.

INTRODUCTION

Since the dawn of time Yoga has occupied a unique and exalted position in India. It suffuses the physical, mental, moral and spiritual culture of the country. It is India's most precious gift and unparalleled legacy to humanity. With the advent of modern sages like Swami Vivekananda and Swami Yogananda, yoga has crossed the shores of India and spread worldwide. Yoga has now become a household word for several millions of people.

Today we live in an age where stress and distress are unavoidable evils. The fast moving modern world has turned life into frenzy, giving it a very artificial frame. The gigantic strides in science and technology, while bringing humanity many advantages and tremendous facility, ease and physical comfort has brought in its wake as many ills. Pollution and its consequences, fast foods with malnutrition and obesity are some of the plagues of modern life.

Modern man has escaped the scourge of infectious diseases to fall into the inferno of non communicable disorders. The plague and cholera epidemics have given way to non contagious, non communicable diseases. Stress related disorders, insomnia, anxiety, fear leading to split personalities, HIV AIDS have taken over the field. Human longevity has been remarkably increased, so have conditions like Alzheimer, dementia, senility and schizophrenia. One person in six is a potential schizophrenic. An indeed bleak picture, considering its demoralising and stigmatising corollaries!

Medical technological progress has allowed the discovery of the causes of disease. Billions are being used by the pharmaceutical industry to discover remedies for ill health. An increasing array of diagnostic techniques is available. Yet, modern science is no closer to giving everybody the chance of enjoying perfect health than it was a century ago.

Man has become a slave of gizmos and gadgets in his eternal quest for peace, happiness, comfort and power. Some believe high style living will bring joy, others feel intense physical activity will bring satisfaction. Many dream of wealth and power to turn their life into heaven. Failing to fulfil his growing desires, man ends up with an eternal dissatisfaction. He loses his peace of mind. As human beings, happiness or unhappiness, health or ill health, depend to a great extent upon the state of mind. That is why the mind is considered as man's best friend or his greatest enemy.

The philosophy of always doing more, faster, heavier etc leads to exhaustion and imbalance. The repeated strain on the system disturbs the homeostatic balance for which one has to pay a heavy price throughout life. Unknown to him, his constant worry and stress are leading man into a state of recurrent homeostatic imbalance. He has forgotten that he has to take care of his health and his being. To make his situation worse, modern man has forgotten how to breathe properly. Being carried away by the flux of modernism he is hardly aware he is breathing. Leave alone that he has to breathe correctly. Yet health and stamina depend primarily on proper breathing.

The greatest flaw of medicine lies in the principle that health of the body is considered apart from the state of the mind or the spirit. Mental tension and increase of mental disease shows failure of these concepts. The cure of disease or repair of injury have become the norms of medicine. Modern medicine has forgotten that prevention of disease or improvement of the individual's life and wellbeing should be part of the equation.

Man is not a mere social animal. He has yet to discover his true nature and the purpose of life. Man is neither body alone, nor is he body and mind. He is body, mind and spirit. To deny this truth will deprive him the privilege of a perfect holistic life. Man will only achieve an exemplary life when his physical, intellectual and spiritual needs are simultaneously fulfilled. As food and sleep are vital physiological needs, yoga & meditation is a vital spiritual necessity for man's mental wellbeing and spiritual unfoldment.

Neither science nor technology has been able to restore the balance. Giant strides in both have only led to greater imbalance, be it in the distribution of wealth, health, security or anywise. The world still suffers from war and starvation. Total annihilation looms over humanity. No harmony in individual life, no harmony in society.

All over the world in every medical school, the concept of homeostasis has been taught for more than a century as one of the basics of health. It is also taught that health is achieved through balance and regulation of our internal systems and that an inability to maintain homeostasis may lead to disease or even death. Yet, achieving homeostasis still appears a far off dream.

The holistic approach in medicine has gone some way to address these problems. Advocates of holistic health believe that health can be achieved through a combination of physical, mental and social well-being only. There is the tendency to forget that we are not just body, nor body and mind but we are body, mind and Spirit. Unless we allow outlet for all three together there will be a dangerous disequilibrium with resulting homeostatic imbalance. A sound body, a sound mind and an eternal Spirit, each relies on the other for perfect coordination and balance. Unless there is perfect equilibrium among all three, man will be faced with ill health, unhappiness and ultimate disaster.

The ideal remedy lies in the holistic approach that yoga and meditation offer. They approach healing and wellness by focusing not just on the body, but on the individual as a whole, his physical, mental, emotional and spiritual needs. One of the reasons that more and more people are turning to yoga and meditation for relief.

The practice of yoga includes proper breathing and control of 'prana', life force or vital energy. Man depends as much on air and prana as essential means of nourishment as food and drink. Air and prana are the basis of life. Impure air and irregular flow of prana in the body will result into sickness and disease.

Life depends entirely on a proper supply of air. We all know that as long as there is air around us we will stay alive. In the absence of air death will follow. Any anomaly in the supply of air would lead to sickness and disease. The ancient yogis having understood its importance have made deep yogic breathing and control of prana the backbone of all yoga practices. Yogic breathing and 'pranayama', control of prana, both aim at normalizing and regulating the proper distribution of oxygen and prana respectively throughout the body.

The control of prana can cure most diseases or mitigate them. Through the practice of pranayama, the supply of prana to a diseased part is intensified, leading to cure of the affected part. Pranayama keeps the flow of prana in a perfect working condition and ensures an adequate supply to different parts of the body. Pranayama restores the homeostatic balance of the body, allowing it to work at optimum capacity.

Unlike the modern physical exercises where quick movements or considerable strain is involved, yoga combines slow, steady, graceful and easy movements. Postures are performed slowly, within the individual's capability, never over-exerting, without the least strain or feeling of fatigue. One is left with a sensation of lightness, peace, relaxation and general wellbeing.

Yoga and meditation are ideal for one and all, young or old, strong or weak, sick or healthy, be it from east or west, north or south. For the healthy, yoga and meditation are preventive. They will help maintain good health. For the sick or ailing, they can restore good health. People with a nervous disposition will find regular practice of postures with its deep, yogic breathing, soothing to their highly strung nerves. Yoga and meditation bring sound sleep without pills.

Ageing is retarded, one gets greater power of concentration and endurance, increased self confidence, tolerance and compassion. The body becomes firm, energetic and active, being rid of muscular, physical and mental tensions. There is ease of movement, enhanced mobility. There is self discipline, greater incentive to rid oneself of bad habits. With growing self confidence, there is greater ability to eliminate nervous tension and improve mental poise. The mind always functions better in a state of calmness.

The aim of meditation and yoga is to keep the human being in a state of optimal homeostasis. Meditation and yoga give us right discipline, the holistic way to restore or maintain our homeostatic balance. We are no more guided by our likes and dislikes but rather according to what is within the norms of proper living, through the practice of proper ethics, proper discrimination, proper conduct and proper nutrition. Yoga and meditation will give us a well balanced, well integrated personality and a strong character.

Homeostasis will not be optimally achieved until we are able to holistically muster all our physical, mental, and spiritual resources to improve our chances of survival, to live happy and fulfilling lives, and to be able to create a utopic society: Yoga and Meditation may well be the answer!

MIRACLE CURES

TESTIMONIES

Yoga and meditation have lost much of their spiritual aura. More and more people are turning to their therapeutic effects in the cure of diseases. For the past few decades extensive research has shown the positive effects of yoga and meditation upon the human system.

Some thirty years back, medical research highlighted the importance of yoga and meditation and dietary changes in a healthy lifestyle. It was successfully demonstrated that heart disease could be reversed by introducing the practice of yoga and meditation and bringing dietetic changes in food habits. It brought yoga the recognition that would open the way to the vast potential of yoga and meditation as therapies for healing and transformation.

At the outset this course was being offered for those interested in spiritual advancement. With time, there were more and more people joining, in search of a possible cure for some longstanding and chronic disease. Many had medical problems for several years. Some cases were considered incurable. Left with no alternative we had to face the reality of allowing everybody a chance. They had all tried different therapies, traditional or alternative, without success. Meditation and yoga with its intrinsic holistic approach cured the majority of cases. The others experienced significant alleviation of disease conditions.

In almost four decades of practice, we have never met anyone who knows how to breathe properly. Everyone thinks breathing is natural and spontaneous. Why on earth do we need to be taught or to learn how to breathe? It may come as a stunning surprise that 99% of people do not know how to breathe correctly or are aware that they are breathing wrongly.

All were surprised when they were taught to breathe properly. Within one or two weeks of deep yogic breathing, most of their complaints whether headache, migraine, insomnia, stress, etc had disappeared.

To quote one among those who got immediate results: "*We were made to understand that the practice of yoga and meditation was more spiritual than anything.*

However, the discipline involved would go a long way to help improve general well-being, bring peace of mind, food habits would be changed, and lastly, yoga postures could remedy physical ills or unknown defects. We were taught yogic breathing and mantra repetition. Within just one week of daily practice I was already another person...Two weeks later when I had my menses it was perfectly painless. I was amazed...After a decade of atrocious suffering I was pain free. It was a miracle!"

She was one among the many who had been diagnosed with infertility. She had been suffering from severe depression consequent to daily family and social harassment.

In some cases, some specialist doctors had even advised hysterectomy, removal of uterus as, according to them, the patients did not have any chances to become mothers. Yoga helped them reestablish homeostasis, regain their health and become happy mothers.

"Within a year of starting yoga, in 1983, we were blessed with a beautiful sturdy baby boy...One year later we were again blessed with a lovely and charming baby girl. Today they are grown up, aged 28 yrs and 26 yrs, and in excellent health...I am in good health up to today and not having any problems. I was so grateful to yoga for having given me back my health... Two of my friends too were blessed with babies. Today their children are grown up, about 22 yrs and 19 yrs old. But for yoga, these two lives would have probably ended in tragedy. Indeed, yoga changed our life and made it worth living and very precious. I wish more people would have the chance I had! How many wasted lives would have been saved!"

There were many other striking cases with amazing results. We have mentioned cases, physical, psychological and mental which cover a wide range of disease conditions. They are some of the thousands who have seen significant changes in their health and life.

To cite, a very serious and advanced case of myelopathy, serious crippling spinal cord disease. The disease could have progressed to quadriplegia, total paralysis below level of the neck. Doctors had advised urgent operative decompression, a very delicate spinal surgery, without any guarantee of success.

Within weeks of doing yoga and meditation his disease process would be arrested. Meditation and yoga have enabled him to lead a normal life these past fourteen years. Let us hear his own words: *"Treating doctors who had predicted complete paralysis of the limbs in the near future, were surprised and perplexed to see my condition improving day by day." (1998)*

"I was examined by a visiting Consultant Neurosurgeon from abroad, Dr P. After

going through my medical records and a physical examination, he shook his head and said: "How is it possible? This guy is stronger than me! His physical examination does not at all corroborate the findings in the records. The MRI shows something quite different!"

"My physical activities are as a normal someone can be. I can walk with all ease several kilometers at a stretch. I still enjoy swimming, driving my car or motorcycle. I do a lot of field work. Today I am 52 years old. Junior officers at work find it difficult to keep pace with me in my daily physical activities. "

There were many cases coming for relief from 'slip disc'. After having followed several treatment regimens to no avail they had given up hope of ever becoming normal again. A few had been advised surgery. They were scared and preferred to try yoga, having heard about its astounding results.

"In 1978...I had badly sprained my back....I had recourse to rubbing balms, deep heat spray or bandaging...I went for physiotherapy, for massage session...I was at a loss. Doctors could not help... I had resigned myself to live with the pain for the rest of my life..."

"I was taught a series of postures, yogic breathing and mantra repetition. My case was already quite complicated and I had to be patient for one or two weeks. ...without realizing, one day I woke up and just got up by myself. I did not need to call anyone to help me up... Unknown to me the back pain was gone. I have carried on with the yoga postures which were meant for a lifetime. Since, I never had any health problems."

To cite another complicated case: *"My neighbour had been suffering from slip disc for some time..... He had to be rushed to the doctor who advised surgery the very next day...."*

"In this condition there was hardly any posture one can do. He was made to perform some standing postures involving the spine and vertebral column, arm and legs. He did the postures with great difficulty....These few postures helped unblock his system. He did not have to go for any operation the next day or ever....He never had any relapse of his problem."

We had several cases of psoriasis. Some longstanding, severe cases had developed suicidal tendencies. They are now cured and leading a normal life. One particularly severe case says: *"I had been suffering from a serious recurrent skin problem since I was 18yrs old. Ever since my initiation and learning of hatha yoga, I have never had any recurrence of the disease, to date, 20 years already."*

We have treated several cases of hyperhidrosis with the best of results. Among the longstanding cases, we quote one example: *"I blushed so intensely that I could feel the blood rushing to my face and at the same time a tension would envelope me culminating with heavy sweating. This phenomenon occurred to me almost everyday and I could do nothing to prevent that from happening..."*

Meditation alleviated those severe symptoms immediately and he was cured within weeks: *"After each meditation session, I feel very cool and relaxed and with a great sense of pride and well being."* Today he is a successful executive.

This serious case was on heavy medication and had been advised ECT, electroshock therapy, by his psychiatrist. Within one week of meditation he gave up all medication (1995). He was cured of all his disorders and is leading a normal life. To date he has never needed any medication:

"I suffered from poor health, a failing memory, and insomnia alternating with somnambulism or hallucinations. I used to have weird nightmares and disturbed sleep... Now, the future holds much promise. I am totally cured when I could have been a case for a mental hospital. Today I enjoy a thriving health, sound sleep and an improved memory. I have undergone a complete metamorphosis. I am a new man with a new destiny! Had I not discovered meditation and yoga 17 years back, I dread to imagine the pitiable conditions I would have been grovelling in!"

Many cases of severe chronic allergy or longstanding sinusitis were successfully treated. Patients were of different age groups. Although refractive to normal medical treatment, they were cured in relatively short periods, some in just a few weeks:

"One of my cousins aged 12, had recurrent sinusitis (1980). Consequently, he was always ailing and underweight. Visits to pediatricians and ENT doctors did not help...He was taught simple breathing exercises and some simple yoga postures suitable for his young age. In no time he was transformed. His sinus problem was completely cured. He recovered his appetite and put on weight. Within six months he became sturdier than his elder brothers." Today, he is tall and well built. He has been symptom free these past twenty years.

Cases of insomnia, hallucinations, chronic headache, migraine, depression, some very severe, used to call by scores. After about 1-3 weeks of meditation they experienced complete relief. Many could have ended as burnt-out cases; some had severe delusional attacks, others had developed alcohol and cigarette addiction. Previous medical and psychiatric treatment was of no avail. Prognosis for recovery was extremely poor. With meditation and yoga their disturbed homeostasis was re-established and they were cured within weeks.

Other cases that have been treated through yoga and meditation include,

Allergy, Anemia, Arthritis, Asthma, Backache, Bronchiectasis, Bronchitis, Depression, Diabetes mellitus, Epilepsy (some severe, longstanding, uncontrolled and refractive to normal treatment), Fibromyalgia, Frustration, Gastric problems, Hypertension, Hysteria, Lack of self confidence, Loneliness, Mental retardation, Obesity, Phobias, Scoliosis, Stammering, Sinusitis, Tendinitis & Tenosynovitis, Urinary Retention, Varicose Veins, Worry and other stress related conditions, among others.

The potential of yoga and meditation as holistic therapies for reestablishing homeostasis and managing or curing different diseases, even severe or life threatening conditions, appears to be unlimited.

These testimonies confirm meditation and yoga as powerful adjuncts in the holistic management of any disease, not only stress linked conditions as is the trend presently.

As someone who has been witness of these 'miraculous' cures states: *"I am always keen to recommend yoga and meditation to everyone. It is the best preventive method and has the virtue of keeping you fit, serene and healthy. It is the best blend of spiritual, mental and physical exercise....If only people had more awareness about yoga and meditation and put them into practice, so many complications would have been averted, so many surgical interventions avoided, so much money saved."*

THE COMPLETE BEING – BODY, MIND & SPIRIT

THE CONCEPT OF HOLISM

Holism means All, Entire or Total. The holistic view stresses that health be viewed from the perspective that humans, as any other living organism, function as complete, integrated units rather than as aggregates of separate or disjointed parts. Consequently, the properties of any given system cannot be defined or explained by the sum of its component parts alone. Rather, the system as a whole determines how the parts behave.

The holistic approach focuses on the causes. In fact, when symptoms develop, it is often long after the causes of the symptoms occurred. The holistic approach in medicine advances that there is more to health than mere symptomatic disease management. Physical health cannot be dissociated from the mental, emotional and spiritual states. The body, mind and spirit are closely interrelated.

Holistic medicine means taking into consideration all facets of the whole person, in the prevention and management of disease. One has to consider the patient physically, psychologically, socially and spiritually. It is rightly believed that there is a relation between physical health and overall 'well-being'. Well-being is a result not only of our physical state in terms of health or disease, but also on its relation with our psychological, emotional, social, spiritual and environmental state. These different aspects should be considered in treating a person truly as a 'whole'.

Modern medical practices focus more on treating symptoms and syndromes. However, an increasing number of physicians are advocating a holistic approach to health care, emphasizing prevention as well as natural treatment. A holistic approach means that the doctor seeks information about a patient's whole background. Maintaining good health should not be limited to taking care of the various singular components that make up the physical body. Other aspects such as emotional and spiritual well-being have to be considered.

Holistic medicine not only treats symptoms but it also looks for any underlying causes of these symptoms. It emphasizes the need to include analysis of physical, nutritional, environmental, emotional, social, spiritual and lifestyle factors.

In truth, medicine was always meant to be 'holistic'. Even the World Health Organization defines health as follows: *"Health is a state of complete physical, mental and social well-being and not merely the absence of disease or infirmity"*. A holistic approach is considered good practice and has been strongly advocated, among others, by the Royal College of General Practitioners for many years.

Holistic therapies have been prevalent from time immemorial. In India, Ayurveda, the 'science of life' has been practiced for more than 7,000 years. In Ayurveda, wellness and disease are considered to be opposite forces where wellness holds disease in balance. One should strive on achieving balance and harmony of body, mind and spirit to maintain health and keep disease away.

The Chinese developed acupuncture some 4,000 years ago and herbal remedies were also common there. North American Indians also had recourse to herbal medicine. Several ancient western healing traditions were involved in holistic medicine. Hippocrates recognized the body's ability to heal itself and cautioned doctors to do their part and wherever possible to allow natural healing follow its own course. Socrates promoted a holistic approach to health. Plato advised doctors to respect the relationship between mind and body.

Holistic medicine focuses on how the physical, mental, emotional, and spiritual elements of the body are interconnected to maintain health. Our body responds to the way we think. Mental conditions affect the body. When one part of the body or mind is not working properly, it has been noticed that the whole person is affected. Holistic approaches focus on the whole person rather than just on the illness or part of the body that is not healthy.

Disease and dysfunction result from a disturbance of the body's harmony and integration. To restore the health of the patient one should seek to restore this lost harmony and wholeness and bring back balance or homeostasis.

HOMEOSTASIS – NATURE'S AMAZING WAY TO PERFECT BALANCE

The human body is among nature's finest masterpieces. This living marvel has the capacity to protect and repair itself. At the same time it has amazing inherent abilities to adjust and adapt to changing internal and external influences. Its different systems work in perfect unison, continuously striving to maintain a state of balance or equilibrium. This ability to maintain equilibrium within is known as Homeostasis.

'Homeostasis' is a combination of two words of Greek origin: '*hómoios*' which means same or similar and '*stásis*' meaning stable or standing in the same place. It is the ability of a living system to maintain a well balanced, same condition; to maintain internal equilibrium by adjusting its physiological processes so as to maintain health and optimal functioning, regardless of external conditions.

Homeostasis is a fundamental property of life and essential for survival. Maintaining homeostasis is absolutely imperative for an organism to stay alive and healthy. Without homeostasis it is not possible for the body to work efficiently and protect itself from harm. To remain healthy, the human body is always striving to achieve, maintain or return to a state of dynamic equilibrium, or homeostasis, both within itself and in relation to its environment. All the systems of the body have inherent regulatory mechanisms that serve to maintain homeostasis.

Homeostasis is one of the most important concepts of physiology and medicine. Both Ayurveda and modern physiology recognize that health is achieved through balance and regulation of the internal systems. According to Ayurveda, the primary cause of disease is imbalance resulting from disruption of homeostasis or immune mechanisms. Hippocrates, the father of modern medicine was inspired from the concept of homeostasis established in Ayurveda, from which he is believed to have drawn much of his inspiration. The modern concept of homeostasis is derived from that of 'milieu intérieur' expounded by Claude Bernard in 1865.

The term homeostasis refers to functional equilibrium in a system or an organism and to the processes that maintain it. Maintenance of a stable constant condition is vital for life. Most bodily functions are aimed at maintaining homeostasis. The inability to maintain it leads to disease and often death. The homeostatic system serves to buffer our body from many external changes and stabilizes our metabolism. Nature has provided the body with multiple regulating mechanisms to make homeostasis possible.

The human body is so designed as to heal itself. This can only occur if it is in a state of homeostasis. The body will thus continuously strive to preserve and maintain this balance. Homeostatic balance is the state wherein body systems are operating within a natural and sustainable range of conditions. These conditions have to be maintained within a very narrow range. Any deviation may bring about disease or even result into death.

The body constantly strives to preserve and control homeostasis to keep the body's internal environment healthy. Even when faced with extreme situations the body strives to preserve or restore balanced conditions that protect it from impairment. Regulation of body temperature is a striking example. In sweltering heat the body sweats to keeps itself cool. Faced with a cold wave the body will shiver to stay warm. In this way it preserves or restores balanced conditions.

In spite of any external fluctuation, core body temperature is maintained between 37.2 to 37.6 degrees Celsius (99.0 to 99.7 degrees Fahrenheit) and is not allowed to fluctuate by more than one degree or so over the course of twenty four hours. If core body temperature goes below 33 degrees Celsius (91 degrees Fahrenheit) a person is liable to die of hypothermia**.** If the temperature goes above 42 degrees Celsius (108 degrees Fahrenheit), death from hyperthermia may occur because cellular proteins are damaged and metabolism stops.

Living cells require certain conditions to survive and function optimally. Homeostasis is the body's capacity to control its inner environment physiologically. Changes occur constantly in and around the cells of living organisms e.g. a change in chemical composition inside or around the cell. Organisms must also be able to withstand external environmental changes. These changes require the cell to react. Homeostasis exists to keep the body in balance, despite fluctuating internal and external environments.

The nervous and endocrine systems are principally involved in the ultimate control over homeostasis as they are responsible for the coordination of the working of all body systems. The central nervous system constantly monitors and immediately responds to changes in the body's conditions. If a parameter strays from physiological limits, receptors detect the change and send signals to the brain. The latter will send a signal to the organ or centre concerned to accelerate or slow down operation. It will trigger a response from the appropriate organ or centre to return the cell and the overall system to the balanced state of homeostasis.

There are two ways to maintain homeostasis; Negative feedback is the most common means for maintaining the body's stability. The other, Positive feedback is less common but does sometimes occur. Feedback is a self-regulating mechanism.

In negative feedback, when a change occurs in the body, it triggers reactions that reverse or negate the change. The brain will send signals to slow down, reduce or shut off. Negative feedback moves to counter the original stimulus. For instance, if the heartbeat increases for any reason whatsoever, the negative feedback will reduce the heart rate and bring it back to a relatively normal rate.

When body temperature is outside normal ranges, the temperature regulating centre is activated. Once the temperature comes back within normal range, the centre is no longer active. If the body is dehydrated or lacking water, a thirst sensation is aroused by signals from the brain. One is driven to drink enough water to quench the thirst. Similarly, when the body lacks food the brain gives a hunger signal, compelling one to look for food.

Positive feedback follows in the same direction as the original stimulus and will result in the cascade effect; more of the same but in greater numbers. For example, in case of injury where bleeding occurs, a clot begins to form. Positive feedback accelerates this process until the clot is able to stop the bleeding. Positive feedback is not as common as negative feedback. At times it does help maintain homeostasis.

Homeostasis is the state in which all the systems within our body are working in perfect harmony. All parameters are within sustainable limits to ensure optimum health, happiness, in short a fulfilled life. Humans can survive and thrive in the most dire conditions. The body has the natural ability and amazing capacity to adjust or adapt itself to changing internal or external environments. It has self-maintaining, self-sustaining, self-regenerating and self-healing properties.

In ideal circumstances, homeostatic control mechanisms should ward off any imbalance. However, under continued pressure, homeostatic imbalance will result. A state of imbalance in the homeostasis of one or more systems invariably reduces the capacity of the body for self repair or growth. The person will be more vulnerable to sickness. If homeostasis is not restored, the health will be impaired. Disease will result. Most diseases can be partly attributed to the presence of homeostatic imbalance within the system.

It is possible for the body to operate under a certain measure of pressure for a certain time, without causing undue harm, for instance working for longer hours or overeating. However, we're stressing all systems of the body. If there is recurring imbalance, damage will ensue with corresponding pathologies. The body will adapt to many changing conditions, but there are limits to these conditions. E.g. In case of a heat wave, we cannot excessively lose water without putting our cells, tissues and organs at risk.

There is a range of responses that is considered normal. Reaching or exceeding these limits can be dangerous. If cells are repeatedly pushed to respond beyond these limits, there may be impairment or loss of normal structure and function. These changes may be reversible to a certain extent only. If they cannot be reversed, the cells will degenerate and disease will follow.

Threats to homeostasis may be of both internal and external origin. For instance, emotional stress resulting from physical or psychological causes, pain or infection and external causes as extremes of weather or external trauma.

Homeostatic imbalance occurs when there is some disturbance in the body's internal environment. This may result from abnormalities in the person's organs and the organ's control systems or feedback mechanisms. The balance of input and output of signals, chemicals, and fluids is thus disturbed. When an organism's homeostatic mechanisms become disturbed, it can lead to diseases.

In some cases, as in ageing, the mechanism becomes less efficient and can no longer efficiently respond to stimuli. This causes an unstable environment that can damage the organism and limit biological processes. Pathologies resulting from homeostatic imbalance include diabetes, hypoglycemia, hyperglycemia, dehydration, gout and conditions caused by toxins in the blood, etc.

Proper homeostasis depends on several factors. Proper diet, proper breathing, proper exercise and relaxation, adequate rest and sleep, a stress free mind are among the most fundamental. A person's diet and stress level are major factors affecting health. The immune system is the worst affected by stress. A fragile immune system is unable to cope with or fight diseases. Dealing with stress by getting adequate sleep, yoga practice, exercising and eating wholesome food will help the body maintain homeostasis and health.

The human body has to face many challenges to its maintenance of homeostasis. A poor diet will make the body compensate or become sick. Drugs, alcohol, tobacco and other toxins may over stimulate the excretory functions in an attempt to prevent these substances from accumulating and damaging the body's cells. Stress and depression can overtax the respiratory, cardiovascular and endocrine systems, thereby weakening their abilities to maintain homeostasis. Inadequate rest or insufficient sleep can exhaust all of the body's systems, impairing the body's balance.

Ideally, homeostatic control mechanisms should prevent this imbalance from occurring. Sometimes the body may not be able to restore the state of homeostasis, as in the case of serious disease, cell mutation, intoxication or malnutrition. When the mechanisms do not work efficiently, or the quantity of the noxious substances exceeds manageable levels, external intervention in the form of medication may become necessary to restore the balance or prevent permanent further damage to the organism. However, medication, whether natural or synthetic, will to some extent hamper the body's natural ability to use its own resources. Although medication may be of assistance, it always interferes with homeostasis. As far as possible, one should have recourse to medication as a last resort.

Constant and excessive stress may have damaging effects on the body, particularly on the cardiovascular system, digestive system and the immune system. There is increased heart rate, high blood pressure and damaged blood vessels. Stress causes the blood vessels to constrict resulting in faulty blood circulation, thickening of blood and defective blood coagulation. Stress increases release of cholesterol into the blood stream. Blood platelets tend to be deposited in the finer arteries further compromising circulation. This will increase the risks of cardiac problems.

During acute stress, blood is sidetracked from the gastrointestinal tract to the muscles, leading to decreased gastrointestinal activity. If the stress is recurrent or prolonged it may lead to stomach ulcers and chronic constipation. Stress will also aggravate existing infections. Although the body tries its best to adjust to constant or higher levels of stress, the resulting strain will weaken the whole organism and lead to illness.

Mental and physical distress upset homeostasis. Disease further increases distress levels which in turn worsen the disease. This never ending vicious circle risks to be fatal, if ignored. With age, homeostatic imbalance becomes more frequent as the body loses ability to adapt to or fight changes in the environment. The depletion of body resources will lead to more serious health problems.

So, while the human body is an amazing entity with superb abilities to counter insults, healthy lifestyles and choices can go a long way to help. It is vital that we respect the natural laws and observe the proper disciplines for a balanced life, whether in food, sleep, work, leisure to maintain the homeostatic balance.

THE TIMELESS SCIENCE

BACKGROUND

The origin of yoga dates as far back as prehistoric times. According to ancient sources, Lord Shiva, one of the Hindu Trinity executed some 8,400,000 postures and gifted them to the human race for the maintenance of health and overall advancement. Among these, some 84 postures are in common use. They are considered as the basis of all other postures.

The ancient sages adopted these postures and practiced them. They taught them to their students. They had a thorough understanding of the physical, physiological, psychological and the spiritual sciences. Later sages would record them in their minutest details as a legacy to mankind. Yoga can be traced back to the Vedas themselves, the oldest Hindu texts.

The first manual of Yoga is believed to be the 'Hiranyagarbha Samhita'. Hiranyagarbha is one name of Lord Brahma, the creator. Some scholars believe that Lord Brahma is the creator of Yoga. Others attribute it to 'Adiyogi' Lord Shiva, believed to be the first yogi. Lord Brahma and Lord Shiva along with Lord Vishnu form the Hindu Trinity.

Famous Vedic sages as Atri, Vasishta, Yajnavalkya and VedaVyasa were among the great teachers of Yoga. These ancient Vedic sages, the seers of truth, were already teaching the Vedas' lofty system of moral philosophy and religion, science and ethics. Their followers were deeply absorbed in the practice of the ethical and spiritual teachings of the Vedas.

Maharishi Patanjali was the first to compile a comprehensive Yoga text, the Yoga Sutras. The practice is known as 'Ashtanga Yoga', the 'Eight steps to Raja Yoga'. The Bhagavad-Gita is the oldest known sacred text on Yoga. The Bhagavad Gita and Yoga Sutras form the theoretical and philosophical base of all yoga. The Hatha Yoga Pradipika is a later authoritative yoga text written by Swami Swatamarama.

Swami Vivekananda's legendary address to the World Parliament of Religions in Chicago on Sept 11, 1893 saw the dawn of a modern era in the history of yoga. The Parliament aimed to lay the foundation for a synthesis of East and West, religion and science. Swami Vivekananda, a blend of eastern culture and western education was the perfect emissary to present the timeless yoga philosophy to an international audience. He laid the foundation for yoga's dissemination worldwide.

In the past century, yoga has seen an exceptional universal popularity. With time its higher values of purity, simplicity and humility have been increasingly overlooked. Its moral and ethical values are being sacrificed to current cultural norms. Many teachers of yoga, ignoring its higher spiritual aims, promote the practice of asanas and pranayamas only. 'Modern yoga' falls short of the ethical and spiritual values the science of Yoga symbolises.

Since the 1960s yoga has become more and more popular in the west. As awareness of its beneficial therapeutic effects increased, it has gained respect and recognition as a valuable method to promote health and well-being. Today many doctors recommend yoga practice for some chronic conditions which find no proper remedy in modern medicine. Among are stress management, heart disease, back pain, arthritis and depression.

Today's life style accentuates stress, tension, worries, seemingly insoluble in the various fields of life. Technology with its great strides has made life more sedentary with all its ill effects on the mental and physical levels. Yoga appears as the ideal remedy. It is day by day gaining importance as the most beneficial therapy for several mental and physical diseases.

Yoga aims at a holistic treatment of a variety of psychological or psychosomatic disorders ranging from sinusitis and asthma to emotional distress. In contrast to allopathic medicine which focuses mainly on the body, yoga takes into account the physical, mental, emotional and spiritual facets of the person.

The combination of yoga and modern medicine is a positive and laudable attempt to link age-old techniques with modern medical knowledge. More medical professionals are having recourse to yoga in their treatment regimen. The move to make modern healthcare more holistic is fast becoming a reality with yoga.

Yoga has an intrinsic holistic approach. It treats the organism as a unified whole, enhancing the physical, mental, emotional, and spiritual aspects of the individual. Yoga postures train and condition the body and mind for a higher purpose. For self improvement, this ageless technique is remarkably efficient. To discover its worth, it only needs to be practiced.

THE ETERNAL TRUTHS

YOGA PHILOSOPHY

If man is the grandest miracle on earth, he is himself his greatest enigma. From time unfathomable, seekers have devoted an entire existence to solve the mystery. They delved deeper and deeper within themselves to unravel the mystery of life.

Man wants to know. He wants to know about himself. What is this existence? Why all this misery and suffering? Man wants to know the truth and free himself. What is the purpose of this life? What is the Self? The worst misery is death. Is there anything beyond death? No one escapes death. King or pauper, saint or sinner, all will die. Is there any meaning to life?

The same basic questions still haunt the human mind. What is the way to happiness? What is the root of human anxiety and suffering? The questions keep cropping up in his mind. Speculation has brought man nothing. Beliefs have not taken him any further. The phenomenal world is always changing. The mind plays havoc. It is also changing constantly. Man wants peace. He still wants to know the truth. He has no control over his life. He is a slave of nature. What is life? Birth, youth, old age, degeneration and death? Everything is imposed on him. While life remains an uncertainty, death is certain. Man is a helpless puppet in the hands of nature. Is there a way out?

The philosophy of yoga proposes the way out. Ancient Hindu seers left humanity the lofty spiritual legacy that is yoga philosophy. These enlightened souls revealed the way out of misery and suffering. They were the first seers and torch bearers in this monumental task.

Yoga philosophy centers on man's suffering, misery and his salvation, physical, mental and spiritual. Man's potential is infinite. The torture of human life is a result of ignorance. Yoga philosophy offers a way to eliminate suffering from human life. By achieving a higher identity he may overcome his miseries.

According to the yoga philosophy, mind, body and spirit are all one and cannot be separated. Yoga is an applied science of the spirit, mind and body. Its practice helps bring about a natural balance of body, mind and spirit, so vital for perfect health and happiness. Yoga creates an internal environment that allows man to achieve an optimum state of dynamic balance.

Yoga is a philosophy that offers insight and guidance into every aspect of human life, the spiritual, mental and physical. It shows that a well balanced, healthy person is someone who is physically, mentally and spiritually harmoniously integrated.

Yoga is the practical part of religion. Man cannot reach truth by reason alone. He has to experience it. He can only experience the truth through the soul, his higher, real Self. The Self can only be known through the medium of the mind. Ignorance is in the mind only. The Self is all knowledge and free.

Everything in time, space and causation is bound. The Soul is beyond all time, all space, all causation. Only when confused with mind and body, it is bound. Freedom from the senses, desires and enjoyment alone will bring man salvation. Yoga teaches that happiness is not in the senses but above the senses and within everyone. Life is only to give man experience, to allow the higher Self in him to manifest itself. Yoga philosophy aims at reuniting our own self with the divinity within.

Yoga is the science that teaches man control of the mind. We have forgotten that we are soul. That we have a mind. We have become bodies. Our aim should be to manifest the soul within. We can only do so through the power of meditation. It is the method that helps man realise the divine in him. Yoga is the path that makes man realise his divinity.

Yoga is more concerned with the meditative side of religion. Deep yogic breathing, concentration, purity of the mind and body are the essential requisites. The ethical side of life should not to be neglected. The mind has to be gradually and systematically brought under control. One has to practice assiduously. It is a life time endeavour, nothing less.

Yoga is the method and the goal. There are various methods to achieve the goal. Our ideals differ. A certain path will suit some better than another. Yoga consists of several paths to suit different tendencies. Among the most popular are Karma yoga, the path of selfless work and duty; Bhakti yoga, through devotion and worship and complete surrender to the Lord; Raja yoga, the realization of divinity through the control of mind; Jnana yoga, through right knowledge and direct experience.

The philosophy of yoga expounds the karmic law. There is no sinful action as such. Man is not a sinner. He suffers because of his ignorance and wrong action. He reaps the consequences of his wrong actions. If he performs good actions he will enjoy the rewards thereof. Man's own actions will free him or enslave him. There is no unjust or just God to punish or to reward anyone.

Karma is not fate. Man has discrimination and freedom of choice and action. He is ruled by reason. He has consciousness. He has vague notion of his subconscious. It is the storehouse of all his past lives' impressions, good or bad. These impressions will surface according to his current actions. Good actions will bring forth good experiences of the past and help him consolidate his goodness. Bad actions will invariably arouse the past bad impressions. These will tend to further degenerate the person. One can mitigate the consequences of wrong actions by doing good.

Religion without philosophy runs into chaos, breeding superstition and fanaticism. Dry philosophy, devoid of religion is nothing better than atheism. Philosophy is theoretical religion. This has to be realised. Yoga is the answer. Yoga and meditation is practical religion. One is the means, the other, the culmination. They are one and the same.

Yoga is probably the only science of religion that can be demonstrated. Yoga teaches us to do virtuous deeds which alone bring knowledge. Knowledge alone can reveal God to us. Neither reasoning nor books can show us God. God can only be realised by superconscious perception. Yoga teaches us how to attain that superconscious state.

Yoga philosophy is ageless. It is the direct experience of countless seers of the past. These have been compiled and form the profound authority of yoga. It is the very core of the spiritual realization of the seers and sages of India. What has been realised by those seers can be experienced by anyone provided one follows the rules.

Yoga is an exact science. It is the science to attaining the aim and end of religion. One does not have to believe in anything. One can experiment for oneself. If it has yielded results once, it will do so over and over again. Yoga is both theoretical and practical. The postures and meditation are the practical side while knowledge of human ills and their remedy is the theoretical.

Yoga is the science of humanity's wellbeing. It deals with the totality of human problems. It is the science that brings forth the humanity and perfection in man.

THE KEY TO LONGEVITY

PROPER BREATHING

A human may survive without food for some three weeks. He may survive without water for a few days. However, if deprived of air he will not survive for more than four minutes. Irreversible brain damage will follow. For all forms of organic life, from the lowest to the highest, respiration is indispensible. Breath is life! Without breath there is no life!

Breathing is the most important biological function of the body. All functions of the human organism are closely linked with breathing. Nature has gifted man with an exceptional respiratory system that can bestow upon him the maximum benefits from life: good health, if not perfect health, happiness, and longevity.

Our health, our mental condition, even our lifespan depend on the way we breathe. Man's life is not measured by the number of years, months or days but by the number of his breaths. The more the breaths per minute the lesser the longevity. Hasty rapid breathing shortens our life. Slow deep breathing ensures longevity. The lesser the number of breaths per minute, the longer, the deeper the breath, the longer will be the life span.

An average person takes 15-18 breaths per minute, each of 3-4 seconds. Most animals take some forty breaths per minute. They have shorter life spans of 12-15 or at most 20 years. Tortoises are slow breathers, taking 4-5 breaths a minute. They are known to be centenarians and can even reach 120-150 years.

Modern man has forgotten how to breathe well. No one cares. We are in the habit of taking things for granted. In these four decades of practice, I have never met anyone who knew how to breathe properly. Mouth breathing and very very shallow breathing seems to be the norm. Suffice to mention that hardly anybody knows how to breathe correctly.

The tragedy is that we mistakenly believe that breathing is not something we have to learn. It is a natural function! We take it for granted we are breathing correctly. Very few are aware of the importance of correct breathing. Hardly anyone among us ever pays any attention to our breathing. In truth it never occurs to anyone, there is a proper way of breathing or breathing is synonymous to life or if breathing stops, life will stop.

Breathing is very essential for the existence of life. It is not an unconscious intake and expulsion of air. Breathing has a twofold objective. During the process of breathing the body gets rid of the stale air in the lungs and replaces it with fresh air. Through inhalation it provides the blood circulation with a regular supply of oxygen. Through exhalation it eliminates carbon dioxide and other waste.

Deep breathing helps maintain an optimal oxygen level, increases blood flow and activates nerve centers. There are certain conditions for proper breathing. Firstly, one should do slow and deep breathing. Secondly, it is a good practice to take more time to breathe out than to breathe in. To expel the maximum stale air, inhalation and exhalation should be to the ratio of 1:2. In deep breathing one breath may last some 15 seconds; inhalation, 5 seconds, exhalation, 10 seconds. Exhalation should be double the inhalation. In this way a maximum of impurities is eliminated.

Doubtless, air is vital for humans. However, breathing is not just inhalation and exhalation. It is one of the most important activities of our body. Proper breathing is of utmost importance since it is the easiest way to absorb *'Prana'*, life force. Prana is the vital force in all forms of life. Without prana there is no life. The air is loaded with prana and we can partake of it through breathing. In shallow breathing only a minimum amount of prana can be absorbed. In deep breathing we can considerably increase our intake of prana, building up a reserve of prana in our body. We will then be able to draw upon this reserve of prana when under pressure or urgent need.

Ancient seers understood the importance of proper regulated breathing to maintain or restore health. Proper breathing should be deep, slow and rhythmical. Yoga teaches us how to use the lungs to their maximum capacity and how to regulate the breath. Deep, slow, yogic breathing lightens the work load on the heart. It also increases vitality and mental capacity.

Oxygen is the most vital nutrient for the body. The brain requires comparatively more oxygen than any other organ. Higher oxygen supply will improve the function of the brain and nervous system. Poor supply of oxygen will cause mental sluggishness, stress and depression. We deny ourselves the chance of perfect health through lack of proper breathing. Our body depends on a plentiful supply of oxygen to rejuvenate itself. Because of poor oxygen supply through wrong breathing, our body degenerates faster. We suffer from premature ageing.

Most people use only a fraction of their lung capacity for breathing. Thus, over 90% of people use just 10% of their lung capacity for breathing and suffer from chronic lack of oxygen. As a result, they feel fatigued, nervous, short tempered and have poor performance. They may suffer from insomnia or sleep disorders. This will obviously give them a bad start the following morning. A vicious cycle! In such cases, the immune system is compromised with loss of resistance. Colds, flu and other infections will be more frequent.

There are four modes of breathing. High, clavicular or collar bone breathing, mid breathing or thoracic breathing, low breathing or abdominal breathing and deep yogic breathing which integrates all three.

Shallow breathing includes the high or 'clavicular' breathing and the mid or 'thoracic' breathing. 'Clavicular breathing' is the most shallow and worst possible type. The shoulders and collarbone are raised during inhalation. Maximum effort is made, but a minimum amount of air and prana enter the lungs. 'Thoracic breathing' is done with the rib muscles expanding the thorax only and is the second type of incomplete breathing. It is more common and comparatively better than the high breathing, but still not proper. A limited amount of air and prana is absorbed.

Low or 'Deep abdominal breathing' is comparatively better than clavicular and thoracic types. It brings more air to the lowest and largest part of the lungs. There is a larger intake of air and prana. Breathing is slow and deep, and proper use is made of the diaphragm, anatomically the main muscle of respiration. The abdomen will naturally expand with the downward movement of the diaphragm. In contrast, in shallow breathing, instead of being expanded the abdomen is contracted during inhalation.

Slow, deep, rhythmic breathing causes a reflex stimulation of the parasympathetic nervous system and decrease in sympathetic nervous system activity. During the process of deep abdominal breathing the stomach, liver, gall bladder, pancreas, spleen and intestines are directly or indirectly massaged and made to function better. With proper breathing it is possible to control or decrease weight and keep it stable. No more pot bellies!

Deep yogic breathing combines all three, beginning with a deep abdominal breath and continuing the inhalation through the thoracic and clavicular areas. The lungs are used to their full capacity. To breathe out, the tummy is first pulled in. Next, the thorax is lowered and last, the collar bones and shoulders are lowered. Both inhalation and exhalation should be carried out in single, continuous, smooth movements. Once the movement is mastered the breathing should be to the ratio of exhalation time twice that of inhalation. Exhalation: Inhalation, 2:1.

Breathing should be full and rhythmical, making use of all the parts of the lungs to increase intake of air. To correct our breathing, we should breathe consciously. In this way we are compelled to breathe more deeply and absorb a larger supply of prana from the air. By far the most important thing about proper breathing is prana. It is important that the lungs be used to their fullest to accumulate a maximum reserve of prana. Control of the prana leads to control of the mind. With regular practice of yogic breathing the mind acquires the capacity to respond to higher vibrations and become superconscious.

For quite some time scientists have known that breathing affects our mental state and our mind. Clinical studies have shown that just breathing exercises can alleviate stress induced conditions like asthma, panic attacks, phobias and irrational fears. Rhythmic slow deep breathing strengthens the respiratory system, soothes the nervous system and calms the mind. The thousand and one desires and cravings decrease. The mind becomes more concentrated, memory improves. Calming the mind and body acts as a natural cure for insomnia.

Deep breathing releases endorphins, natural painkillers found in our body. This can ease muscular tensions, relieve headaches, sleeplessness, backaches and other stress associated aches and pains. Deep breathing can also partially compensate for lack of exercise and inactivity following some illness or injury. It also helps recover faster from stress and exertion.

Shallow breaths may allow man to survive but he does not realise the great damage it is causing to his health and natural growth. From time immemorial the ancient yogis of India have warned against the ills of shallow breathing. They taught humanity the wisdom of the science of breathing. Poor breathing causes early death, while its control will extend life.

Modern science has recognized the dangers of shallow breathing. Shallow breathing is responsible for many ailments, like heartburn, chest pain, palpitation, stomach disorders, muscle cramps, dizziness, anxiety, sleepiness, visual problems and fatigue, among others. There have even been cases where people apparently having symptoms of heart disease were in fact suffering from consequences of their poor breathing habits.

Today a high percentage of people suffer from various breathing problems. Either they do chest breathing or shallow clavicular breathing. What is worse, some 50% of people are mouth breathers. Mouth breathing is responsible for aggravating many chronic diseases. In children mouth breathing may give rise to frequent infection, asthma, dental caries etc. In adults, mouth breathing affects general health and may give rise to morning fatigue, headache, snoring, sleep apnea and dry mouth syndrome, among others. The mouth is meant for eating and drinking, certainly not for breathing. It would be an absurdity if someone tries to eat through the nose. It is as absurd for anyone to breathe through the mouth in normal circumstances.

Nature has provided the nasal passages with exquisite mechanisms so that neither cold air nor dust particles can penetrate our system and endanger health. One should always breathe through the nostrils as this will filter and remove impurities from the air before it reaches the lungs. The nasal mucous membranes and the tiny hairs in the lining of the nasal passages will trap dust particles and kill bacteria and viruses.

Cold air which may disturb our homeostasis will be warmed to body temperature. During winter, inhaling icy cold air through the mouth can lead to constriction of the coronary arteries. It may in turn trigger a heart attack in those susceptible. The olfactory system is another marvelous apparatus to detect and warn of noxious matter. Mouth breathing does not provide us with all this protection. That is why it is imperative that we breathe through the nostrils, as nature demands.

The power of proper breathing lies within us. Unfortunately, some 99% of humans do not know how to breathe properly. As a result, oxygen levels in the body cells and prana levels are drastically low. The best remedy to this appalling situation would be for those who have knowledge of proper breathing to help others at home, at school, at work or anywhere to breathe the proper healthy way.

Breathing affects the entire human organism. It reduces mental and physical fatigue. This single practice brings us both vigour and relaxation. Even those with no health problems confirm that proper breathing makes them look better, feel more relaxed, have greater output and more zest for life.

Correct breathing is perhaps the easiest and most important technique for safeguarding health. Knowledge of proper breathing techniques would enhance health and quality of life through increased vitality and rejuvenation.

DEEP YOGIC BREATHING

Sit upright on the floor. In the beginning it is not necessary that one should sit in lotus posture. Easy posture, Thunderbolt posture or Warrior posture are as suitable until one can master the lotus posture. The spine should be kept straight, not to obstruct the abdomen, thorax and the collar bones, the three parts of the body used in yogic breathing. Breathing should always be done slowly and smoothly, through the nostrils. Never breathe through the mouth.

Prior to practicing all three phases, it is easier to start with 'low' tummy breathing. Place one hand on the tummy and the other on the chest. Only the hand on the tummy should move. Follow the in and out movement of the tummy. Tummy out when inhaling. Tummy in while exhaling. The hand on the chest should not move.

Close the eyes and concentrate on the repetition of some sacred syllable as 'OM'. When breathing in, mentally repeat 'OM' once. When breathing out mentally repeat 'OM' twice. The ratio of Breathing in: Breathing out should be 1:2.

To ensure correct breathing always remember this mnemonic: BREATHING IN, TUMMY OUT. BREATHING OUT, TUMMY IN!

METHOD

A. Low breathing

- Breathe out first, pulling in the tummy.
- Breathe in slowly through the nostrils, drawing in maximum air.
- Breathe out slowly, pulling tummy in, expelling air out of the lungs through the nostrils. Only the tummy moves while the chest remains motionless.

B. Middle breathing.

- Breathe out.
- Breathe in slowly through the nostrils while expanding chest on both sides.
- Breathe out while contracting the ribs, expelling out air through the nostrils.
- Only the chest moves, while tummy and shoulders remain motionless.

C. Upper breathing

- Breathe out.
- Breathe in slowly through nostrils raising collar bones and shoulders.
- Breathe out slowly, lower shoulders and expel out air through the nose.
- In upper breathing the abdomen and chest remain motionless.

D. Complete yogic breathing

- Once tummy breathing is mastered, the complete yogic breathing can be practiced with ease. A complete yogic breath starts with a deep abdominal breath and continuing the inhalation through the thoracic and unto the clavicular areas. To breathe out, the tummy is first pulled in. Next, the thorax is lowered and last, the collar bones and shoulders are lowered. Inhalation and exhalation should be carried out in single, continuous, smooth movements. Allow a short pause between inhalation and exhalation. Once the movement is mastered the breathing should be to the ratio of 2:1; Exhalation 2: Inhalation 1.

The highly strung nerves will calm down. One will feel genuinely rested for the first time. Coupled with yoga and meditation, deep yogic breathing will go a long way to transform our life. It is neither difficult nor dangerous. The more it is practiced, the more there is calmness of body and mind.

PRANA – THE DRIVING POWER OF THE UNIVERSE

PRANA – LIFE FORCE

Prana is the primordial force, the universal principle of energy. Prana is the infinite, omnipresent power of the cosmos. The whole universe is charged with prana.

Of Sanskrit origin, the word *'Prana'* means life force or vital energy. It is the latent energy out of which all life, all forces, all activities are born. Nothing exists without prana. Wherever there is life or activity in the universe, there is prana. All living organisms from the lowest being to the highest depend on prana to keep alive. Prana is life. Without prana there is no life. Prana is the basis of all life and energy. Prana is the driving power of the universe and is in every manifestation of life.

Any activity whether in the physical, mental or spiritual world is the result of vibration. That vibration is the result of prana. Out of prana is evolved everything that we call force or energy. Every force, physical, mental or spiritual has as basis, prana. Prana manifests as motion, gravity, electricity, magnetism, etc. Invisible, immeasurable and indestructible, prana permeates every iota of life.

In all forms of life, prana is present as the 'vital energy'. Prana is the cosmic life force that brings inanimate matter to vibrate with life. It is prana that causes the tiny seed to germinate and grow into a huge tree. The magnificent rose blooms thanks to prana. It is the same prana that causes breath, motion in the muscles and makes the blood circulate in our body.

Behind all actions of the body there is prana. The nerve currents and the thought forms are prana. Prana is present as life force or vital energy, to protect, maintain or restore equilibrium in the natural growth or unfoldment of life. Prana nourishes both mind and body. The mind and prana have a special relationship. Prana is intelligence. The mind makes use of prana to express itself in thought. Thought is the highest and finest action of prana; instinct or unconscious thought giving rise to reflex actions is the lowest action of prana. From the highest down to the lowest, every force is but the manifestation of prana.

Prana is plentiful in the air around us. We are bathed in an ocean of prana. The most obvious manifestation of prana in the body is in the motion of the lungs. The air that passes into our lungs through breathing possesses that prana. Should that stop, all other manifestations of force in the body will come to an end. Prana is also absorbed through food and water. Wholesome food, clean water, fresh fruits, vegetables and fresh milk contain an abundance of prana. Sunlight is another source of prana.

There is an intimate connection between life and breath. Without breathing there is no life. To survive we draw prana from the air we breathe. Breathing is the most important biological function of the organism. Every other activity of the body is closely connected with breathing. Breathing is vital for our health, our emotional balance and well being and even for longevity. That is why proper breathing is exceedingly essential.

During respiration our body absorbs prana from inhaled air. In shallow breathing our intake of prana will be low. The lungs working at a fraction of their capacity will allow a limited quantity of prana to be absorbed. The more the prana we can draw from the atmosphere into our system, the greater the chance for us to enjoy optimum health and have reserve prana to fight infection, disease or some other demand.

Proper breathing is the fastest and easiest way to absorb prana. Human beings expose themselves unnecessarily to sickness and disease by haphazard, shallow breathing. Deep breathing uses the lungs to maximum capacity. There is an abundant inflow and absorption of prana. The body is able to build up considerable reserve of prana to help develop enhanced vitality and face any emergency. If deep yogic breathing is practiced until it becomes second nature, there will be an abundance of reserve prana in the organism.

Prana is that force which keeps the body alive and maintains a state of good health. If there is insufficient prana in our system, the body will degenerate prematurely, as disease develops. Every living form has the power to self repair. Prana bestows this capacity to the living organism. An abundance of prana helps the body protect, preserve and enhance its power of self repair. It also helps develop resistance to all negative influences within or without.

There are people who endure even the worst of hardships, yet never collapse. They always display amazing resistance. They are physically and mentally strong and possess remarkable vitality. It is nothing unusual. Knowingly or unknowingly they have tremendous prana reserve. They breathe the proper way, filling their lungs to full capacity, drawing prana from the cosmic bounty around. They are thus able to store more prana. The more prana that is stored up the stronger the individual, physically and mentally.

Every living organism is exposed to threats or changes in their internal and external environment. In order to adapt to this ever changing environment the organism needs a certain amount of energy. Prana is that vital energy which sustains all living things in their every stage of growth and development. No disease or harm can affect the body if prana is being constantly replenished and there is reserve prana in plenty. Prana will help the organism destroy all germs, bacteria and viruses entering our body through breath, food, drink or through the skin and try to attack our tissues and cells.

If prana is unwisely overtaxed through misuse, waste or dissipation, through lack of proper food, through violating of natural laws, health will suffer. Adaptation to the environment will be faulty. The prana is unable to fight back or adapt itself to the compromised conditions. Health is impaired. The body becomes weak and vulnerable, unable to fight back any attacks. It becomes more prone to various disease, sickness or ill health.

Prana instills self preservation in every organ and tissue. Even the minutest component of the organism, every single cell, uses this power to protect and defend itself and to heal itself. In case of threat, each cell will immediately react to get rid of the invader and safeguard its natural physiological state. If some cells are attacked by some germs, they are weakened. Other cells imbued with an abundance of prana will come to the rescue and will fight against this enemy to ensure that normal conditions of the body are restored. If the body's defenses fail to protect it, pain, aches, disease and even death may follow.

Contrary to the common belief that drugs and medicines have the capacity to heal, it is in fact the body that heals itself. It is true that medication will protect the body from impending infection or from further deterioration, but the body will ultimately heal of its own. Medication is administered to help this process by temporarily alleviating pain or decreasing inflammation or infection. In the case of a bone fracture, the limbs will be put into a plaster cast for immobilization and proper repositioning only. The bone will heal on its own, naturally.

In fact, 'green bone fractures' in children are usually allowed to heal in their own time without any surgical intervention because healthy children have an abundance of prana. Similarly, all plants and animals are steeped with prana and its healing power. Broken branches will re-grow. Animals in nature are exposed to injuries and illness, but they do keep very healthy. Closeness to nature allows their organism to be permeated with prana to protect and heal them.

Wholesome natural foods should be preferred to junk or fast food. Under pressure, the digestive system will try its best to assimilate these. However, sooner or later this constant abuse will result in indigestion, constipation or diarrhea, inflammation of the guts or even cancer. Drinking impure water will invariably introduce germs and parasites into our body, with catastrophic results. When there is a change of season or external temperature fluctuations we may be subject to heat stroke, colds, influenza, allergy etc if we do not take timely precautions. Doubtless the organism will try its best to neutralize these adverse circumstances, but precious prana would be wasted. Recurrent emergency or prolonged stress will diminish prana reserve and weaken the organism's resistance.

The quantity of prana varies according to its use and abuse or misuse by the individual. If prana is naturally engaged, it will maintain the system in perfect health and harmony and help the body recover from any unforeseen attack. If pressed to work in adverse conditions it will become severely impaired and inordinately weakened. The whole system may collapse. Sudden bouts of emotion, a sudden wave of anger, fear or violence, abuse of sex and such conditions where the body's balance is imperiled will cause prana to dwindle. The body's homeostasis is disrupted and it becomes extremely vulnerable to attacks, internal or external, as in the case of tuberculosis, AIDS or cancer. Prana in this deplorable situation is unable to function normally. The organism will gradually or rapidly ebb away.

Our health is dependent on the natural, balanced flow of prana within the body. Prana maintains harmony in our system. Perfect harmony is synonymous to perfect homeostasis. Perfect homeostasis is perfect health.

PRANAYAMA – CONTROL OF THE LIFE FORCE

The word *'Pranayama'* is of Sanskrit origin, *'prana'* means 'life force' or 'vital energy'; *'ayama'* means regulation, extension, amplitude etc. The knowledge and means to control prana is the science of pranayama.

Prana is vital for life. It is the sum total of the cosmic energy that is present in every organism. It keeps the body alive and in perfect shape. Prana works in our mind and body as the mental and physical energies. Pranayama means the control of prana.

We have to try the control of prana from wherever it is within our reach. The most obvious manifestation of the prana in our body is in the motion of the lungs associated with breath. Breathing is probably the easiest and most accessible method to control prana. Prana is invisible but on the physical plane it manifests causing inhalation and exhalation, producing breath. Prana is not breath. Breath does not produce the motion of the lungs. On the contrary, it is the movement of the lungs that produces breath.

The first step in the control of prana is to control the motion of the lungs. Pranayama is not breath control, but control of the prana which causes the movements of the lungs. The motion of the lungs has intimate relation to the activity of the mind. If the mind is controlled the motion of the lungs may be controlled. Vice versa, if the motion of the lungs be controlled, the mind can be more easily controlled. Breathing is one of the body's few functions which can, to a great extent, be controlled both consciously and unconsciously. Controlling the breathing motion is pranayama.

The respiratory center which regulates the respiratory system has a certain control over all the other systems of the body. If the motion of the lungs be brought under perfect control, it would be easier to have control over every part of the body. Breathing at a certain rhythm tends to bring a corresponding rhythmic action in the body. Pranayama brings about regulation of the breathing and thus, rhythmic action of prana. This rhythmic action of prana will help us gain control over the whole body. When prana works rhythmically, good health prevails.

The aim of pranayama is to bring harmony in the bodily functions and restore the perfect balance, homeostasis, essential for perfect health. Disease is caused by an imbalance of prana rhythm in the body. Any sickness or disease will indicate that prana is deficient somewhere in the system. When the balance is restored by directing a fresh supply of prana to the affected part, the rhythmic action of prana is restored. Healing will then follow. This transfer of prana from one part of the body to another is another form of pranayama.

The life force, prana is the prime mover of the mind. To control this prana is to control the mind. Body and mind act upon each other. Every mental state results in a corresponding state in the body. Any change in the mental function produces corresponding physical changes in the body. Similarly, any bodily change will have corresponding effect on the mind. Modern medicine has recognized that many diseases, physical, mental or psychosomatic have as root cause the mind, through wrong ways of thinking.

The mind and the breath are closely interconnected. An uneven and rapid breath causes distraction of the mind. Similarly, an overactive or disturbed mind causes irregularities of breath. By controlling the prana, perfect control over both the body and the mind may be achieved. Thought is the finest and highest action of prana. Mental suggestion produces some wonderful effects on the body. The latent powers of the mind, if developed, can cure ailments or disease.

To enhance the supply of prana and ensure its proper distribution to every part of the body is the science of pranayama. When the prana balance is disturbed, sickness or disease will result. A redistribution of the prana to restore the balance will cure the disease. Every part of the body can be filled with prana and when it is so the whole body can be healed. All disease or fault in the body can be easily controlled. To know where there is less or more prana and balancing the prana in the whole system to maintain perfect health is also pranayama.

Since time immemorial, a long life with perfect health has been considered the greatest blessing of earthly existence. Perfect health is not merely strength or vitality of the physical body. It is the homeostatic state of equilibrium of the vital currents, subtle energies and all the organic functions. It is the harmony between internal nature and external nature, between the mental and physical bodies. This can be achieved through the practice of pranayama.

Achieving and maintaining good health implies more than just care of the various parts of the physical body. One has to understand the laws that govern our life and follow them. The body will then have an abundance of vital energy, prana, to destroy all disease. Reserve prana will fight back in case of any emergency and help maintain perfect health. For this purpose the ancient seers advocated the practice of this wonderful science of pranayama.

Yoga is the method to control prana as a mental force. The ultimate goal of yoga and pranayama is to control and harmonise the flow of prana all over the body. When prana is well balanced, healing power is acquired. Pranayama helps maintain a state of perfect homeostasis and health.

Deep yogic breathing is the simplest and easiest method to control prana. It is suitable for everyone, healthy or sick, young and old alike. It is the only breathing practice taught throughout our yoga course and has brought excellent results over the years. The importance and value of correct breathing is to save and enhance the vital energy or life force, prana.

The advanced methods of pranayama are for acquiring higher powers. Most may be quite hazardous and should be practiced only under expert supervision. Complicated, difficult methods of pranayama also require appropriate diet. Else, there may result depletion and damage of the system which may lead to imbalance and disease.

To derive the maximum benefits from pranayama, we have to keep in good health through a proper diet, appropriate exercise, adequate relaxation and sleep, and by avoiding stress to the maximum. Pranayama calms and purifies the body and the mind. The high-strung nerves will calm down. Then one discovers what rest truly means. Never before would one have known such rest. Pranayama will help acquire physical health, mental peace and purity, intellectual acumen and spiritual fulfillment.

DON'T EAT YOUR LIFE AWAY

DIET

We are what we eat. Out of the food we consume, our body manufactures everything required for survival. Our food is not only concerned with physical growth, but it profoundly affects our mind. We depend on the food we eat for prana too. This vital energy plays a crucial role in maintaining homeostasis, for us to enjoy optimum health.

The aim in planning a proper balanced diet is to provide the body with maximum energy for strength, protection and longevity. It should also provide us with mental strength and spiritual energy. We should only choose such food that keeps the body healthy and light, the mind calm, alert and pure, and enhance our immune system.

A proper diet should cater for body and mind efficiency through such food that will provide us with an abundance of prana. Natural fresh foods contain by far the maximum of prana. Fruits, vegetables, seeds, nuts and grains are those that will bring us the most prana, besides other essential nutrients.

The physiological meaning of food is assimilation of energy from the sun. The sun supplies energy to all life on earth. Solar energy is full of prana. By eating fruits and vegetables we ensure ourselves of direct sun energy in them. Plants grow by taking in energy direct from the sun. Herbivorous animals feed on plants. Carnivorous animals prey on the latter. As we move higher in the food chain, the lesser the prana there is. Animal flesh is a secondary source of prana.

Balanced diet cautions us to moderate our intake of food, whether in terms of size or number of meals. Rather than taking a few heavy meals we should take light meals more often. We should adjust our food according to our health needs. Whatever existing excess, should be eliminated gradually. The yogic method recommends one to proceed in a slow systematic manner. One should not try to drastically alter anything. Any sudden cut may cause an imbalance because the body cannot readjust fast enough. It may cause unnecessary stress to the system.

Many diseases are caused by irregularity of meals, overeating or taking unwholesome food. We should develop the habit of eating only when hungry, at set times. We should not stuff ourselves but rather end a meal with a feeling of some empty space in our stomach. Sometimes stress, some mental or bodily upset incite us to nibble at anything we can lay hands upon, a way to bridge over the gnawing hollowness. There is no enjoyment. It is mechanical. Worse, we are not even aware how much stuff has been ingested. It is certainly a bad habit. The earlier we put an end to it, the better.

Food plays a vital part in keeping up perfect health. The best choice is to take food that is nutritious, filling and easily digested. We should choose such food that in small amount offers the best nutrient. They are fast digested and assimilated using the minimum energy. Too much energy wasted in digestion will leave less for the other functions of the body. Overeating will produce a feeling of lethargy. Simple meals do not excite or over stimulate either our body or mind. There is easier digestion and assimilation.

A proper and balanced diet consists of adequate supply of proteins, carbohydrates, fats, vitamins, minerals, trace elements, water and roughage. Fruits and vegetables, preferably raw or half cooked, whole grains, nuts, germinated seeds, dairy products, honey, dry fruits are among the best. Vegetables should never be overcooked as they lose most of their nutritive value. Fruits should be eaten fresh and as far as possible, raw.

Because of the rise in environmental pollution, we all need a diet rich in antioxidants to neutralize all the toxins present in the environment. Vegetables, fruits, whole grains, legumes, and nuts are of prime importance because of their high antioxidant content. Certain complex antioxidants like flavonoids are only present in vegetables and plants. Eating a variety of fruits and vegetables will protect us and help us fight the adverse effects of modern life. Recent research has shown that vegetarians enjoy better health than non vegetarians.

Proteins from nuts, dairy products and legumes are generally of a healthier quality. Mixed pulses make an excellent source of proteins. Whole wheat flour, rice, barley, milk, sugar, honey are good staple foods. White flour, highly processed, has fewer nutrients. It may be mixed with gram flour, whole wheat flour, oats or semolina.

A surplus of proteins, carbohydrates and fats are risk factors for disease. Modern busy lifestyles makes many turn to fast foods which usually contain these in excess.

This explains the epidemic of non communicable diseases such as diabetes, hypertension, cardiac problems etc. Recent research has shown that communities who are giving up traditional eating habits and relinquishing homemade food for fast food are increasingly at risk of so called 'life style' diseases. Non communicable diseases are fast growing in these communities as diet habits are changing.

Food should be freshly cooked, not stale nor re-warmed. Commercially prepared food from restaurants, hotels, take away, junk or fast food usually cause disease. The food may be stale or polluted. Food sold in the open, by the roadside, collect dirt and dust and is exposed to flies and pests. Many illnesses result from buying impure food from the market.

One should shun exciting food such that are highly spiced, sharp, sour, pungent; hot foods, like mustard, liquors, fish and meat. One should avoid any artificial stimulants, alcoholic drinks, excess of salt and acid or anything difficult to digest. Excessive coffee drinking is a serious hazard to health and may give rise to nervous problems.

A sound mind in a sound body is man's greatest wealth. Food has great influence not only over our physical well being, but also over our thoughts, our emotional and spiritual well being. A poor unbalanced diet weakens the body and adversely affects the mind through lack of prana. Wholesome and nourishing food with an abundance of prana is essential to maintain health, vigour, strength and vitality.

One should eat to live a long life, rather than live to eat one's life away!

WIPE OUT LIFE'S SUFFERINGS

ETHICS

Happiness and misery are two different conditions of our mind. They are of our own making. It is our own ignorance that is at the root of all our miseries and suffering. The world is neither good nor evil. We make it good or evil by our perception and our tendency to depict it with optimism or pessimism. Take for instance the well known expression: Is the glass half full or half empty? There may be areas of our life where our thinking needs to be reassessed. Positive attitudes will go a long way to turn our miseries into a blooming happy life.

The negative tendencies of the mind can be attenuated, if not altogether neutralized by cultivating positive habits. Violence, arrogance, anger by love, tolerance, courtesy; vanity, boastfulness by humility; greed, attachment by contentment and detachment; abuse or misuse of sex by continence and abstinence; dishonesty by compassion and truthfulness; jealousy and envy by sharing and acceptance; hatred by friendship and appreciation, and so on.

When good preponderates in the individual he will progress and help others progress. Should evil predominate, man degenerates and brings misery to himself and ruin to others. The law of the jungle will prevail where morality sinks into oblivion. It will be survival of the fittest if the brute in man is aroused and nurtured through violence or non respect of others' legitimate rights. If the animal instinct in man is let loose, only the brute will be left, turning humanity into a horde of unthinking, violent and vicious criminals. Crime, rape and theft will be the norm of the day. People will live in permanent fear and anguish, leading many to suicide.

Wherever ethics is taught as part of living, it serves as a beacon for man. A life devoid of ethics is an existence devoid of beauty, peace and tranquility; without harmony and happiness; without feelings and decency. Without moral values, our life will be scattered, obsessive tendencies plaguing our disturbed mind, with only negative repercussions to harvest as we tumble downhill to our own destruction.

The ethical laws should govern human nature. They will teach man how to overcome evil with good. Violence can never be stilled with violence. Nor can hatred be wiped out by hatred. It can only aggravate the situation, where all will be losers. An eye for an eye will turn everyone blind. Ethical laws teach us to forsake all evil and bring forth good. There is nothing higher than overcoming evil with good. It is the only path to progress and perfection.

Sound moral values help to counterbalance pronounced evil, vicious tendencies and mould man into the finest jewel of creation. Our daily life if guided by these higher values will spare us the endless sufferings, disappointments and miseries. Rather than live a life of wanton enjoyment and pleasure and reap all the ills, we should learn to enjoy life in a disciplined way. All abuse will eventually harm. Whatever the abuse it always degenerates man whether physically, mentally or morally.

All ethics, all moral virtues are of divine origin. They are meant for man to curb his animal instincts and become a monument of virtue with an exemplary life, fit for a god. Those great virtues are to be daily practiced. All these do's and don'ts boil down to not being selfish. The true basis of morality is unselfishness. All the regulations and proscriptions found in different scriptures are more to curb selfishness, the great bane of humanity.

Selfishness is the root cause of all evil and immorality. Selfishness prevents us to recognize the right of others. We do not mind using fair or foul means to better our lot, injuring others or depriving them of their rights. Our mistake is that we forget we all belong to a same world. The world belongs equally to all of us. We should live in solidarity rather than hostility. All selfish actions are immoral and sinful. That which leads to unselfishness is moral, virtuous and good.

The grandest of the ethical laws as taught by the Vedanta is: *"Love every living creature as thyself"*– nothing more, nothing less. It is the law of universal love. It includes not only humans but the lowest of animals, because the same Spirit or Self resides in all. He who recognizes that the One and same universal Spirit is everywhere will refuse to harm anyone.

Any action that makes us unselfish, acts as a purifying balm. Any action that uplifts us from our lower instincts draws us closer to the Supreme. We should help and share in fellow beings' misfortune. Relieve them from pain and suffering.

We should love and share with all creatures as one great entity. Until this oneness is realized and the idea of separateness is dropped, the tendency to mine and thine will persist. Selfishness will bare its ugly fangs.

Ethics are designed to help man conquer his animal instinct. His greatest enemies are his violence, anger, lust, attachment, his vanity and envy that overpower him. Moral disciplines ennoble man's interior life and help root out the secret, endless desires surging within the mind. These disciplines alone can counter or altogether neutralize those negativities that make life so unworthy of its name.

Negative emotions create stress. Stress invariably leads to tension, the source of many diseases : anxiety, despair, anger, headache, depression, frustration, chronic fatigue, loss of memory, mental imbalance, high BP, palpitation, chest pain, heart attacks, etc. Chronic anger, worry and hatred increase the risk of disease. In contrast, positive emotions result in better health, longer life and a great sense of well being.

There is a perpetual inner battle within us to resist temptation. Ethics teaches us not to allow temptation to tempt us. Once we give in to temptation, it will be very difficult to stop. It the first step down the precipice that seals our eventual downfall.

We should always bear in mind that all ethics, all ethical teachings, moral ideas, moral and spiritual laws are for curbing the strong attachment to our lower instincts. They simply urge us to take the opposite path, away from these negativities. All crimes are as a result of our desire for possession and physical pleasures. Many are so avid of enjoyment. Their thousand and one desires put their mind in permanent turmoil. They have no peace, no respite and their health takes the toll.

All competition, struggles and evils that plague our life are manmade. They can therefore be remedied through team spirit, sympathy and sharing. The oneness we feel as a team is the basis of all ethics and morality. This spirit of solidarity and unity yields finer results. The aim of ethics is this unity, this sameness. We may not be born equal. One is born wealthy and another in poverty; one strong, another weak; one is intelligent, another less so. The aim of ethics is to bridge the gap between these eternal disparities by providing equal chances to all.

Morality is not an end in itself, but a means to an end. There is only one way to define morality. That which is selfish is immoral and that which is unselfish is moral. Every selfish action takes us to our downfall. Every unselfish action uplifts. Unselfishness, the foundation of all morality and the quintessence of all ethics, should be the basis of our life.

The whole scope of morality is to prevent man from degrading himself to thorough selfishness. Morality consists of both restraints and the cultivation of certain virtues or higher qualities. These qualities serve as antidote to man's negative tendencies. They also serve as a catalyst to show him the way to genuine humanity.

Morality and purity are the only forces to destroy ignorance and evil. Ignorance and evil destroy humanity. They need to be destroyed by cultivating goodness and awareness. We have to become moral to discover true happiness. Until we are moral we cannot wipe out life's sufferings.

YAMAS & NIYAMAS

The Yamas and Niyamas are of divine origin. 'Yamas' are moral restraints and 'Niyamas', moral disciplines. Two simple words on which depends the fate of humanity! Either man does justice to his Divine image and become moral. Or miserably gropes in the darkness of ignorance and falls into the quagmire of immorality.

Too often yamas and niyamas are mistaken as exclusive to the practice of yoga and meditation. Nothing can be more misleading and further from the truth. Ethics as universal moral laws are fundamental to every human being, man, woman and child, young or old, rich or poor. Wherever one may be, whoever one may be, all are concerned where health, happiness and harmony are at stake. Nothing compares with man's discriminative faculty. There is nothing superior to man's intelligence. He has the potential either to uplift himself through moral living or degrade himself through immoral living.

Restraints (Yamas)

1. Non violence

Non violence is not just non-killing of humans or animals. It also implies refraining from injuring them whether by words such as abuses, accusations, insults, harsh words; in thoughts, by harboring hatred, evil thoughts and revenge against anyone or by deeds, physical assault. Even intent to injure is a violation of the principle of non violence. Treat with kindness and compassion humans, animals and plants. Life is sacred. Respect life in whatever form. We are all part of a whole, interconnected and interdependent.

Replace hatred by friendship; revenge, hostility, anger by forgiveness. It helps to promote peace, trust, love and friendship and removes fear, mistrust, enmity or hostility. Feelings of hatred or ill will are neutralized, protecting the mind from undue stress. It relieves tension, reducing the risk of a great number of stress related diseases. It is excellent for a healthier and happier life.

Non violence also means mutual consideration and sharing. Settle differences through recognition of one another's rights. Reconcile differences through compromise. Only then will enmity or stress not overpower the mind, body and emotions, leading one to commit injurious acts. Positive attitude begets positive attitude. It is a way of protection for each and everyone.

Good thoughts mean a friendly disposition. It will eliminate fear of the other and the resulting stress. Both mentally and physically we will be stronger. In the long run the world will be a better place to live where peace, love, trust reign supreme; nothing short of heaven.

Neuroscience has shown that violence results in abnormal brain development. This inhibits the person's ability to develop humour, empathy and attachment, to counteract the effects of violence. They instead develop other ways to deal psychologically with it. Men have been known to develop a 'fight or flight' response. They become prone to alcohol, drugs or other addictions, which may lead to further violent behavior. Women and children are more likely to recoil into themselves. Many suffer from depression, amnesia, suicidal tendencies and other mental illness.

It is commonly said that if one perfects the practice of non violence, there is no need for any other practice, for all the other practices are intrinsically linked with it.

2. Truthfulness

Truthfulness is among the highest virtues. Falsehood is an ugly sin. Whatever the situation, we should never tell lies. Lies pollute the mind and can only hurt the liar in the long run. We should always be truthful in all circumstances.

Although quite difficult to practice, yet to instill confidence and trust in others, we have to be truthful. We should not hurt or cause harm to others by telling lies. It is as sinful to encourage others to tell lies or approving others' lies.

Close ties should share all secrets, pleasant or unpleasant. They are the ones that will share in sorrow or happiness or bring us the best advice. Never betray trust, secrets or promises.

3. Non stealing

One should never appropriate others' property or belongings. Give up unnecessary necessities or cravings by cultivating self sufficiency.

Excessive desires for things we cannot possess might lead us to steal. Whatever we cannot obtain, we will be tempted to steal. We will try to acquire them by fair or foul means, transgressing others' rights.

Wanton wastage of personal possessions, others' property or natural resources is equivalent to stealing.

Non stealing teaches us to be satisfied with what we can acquire by honest means, not by stealing or accepting free gifts and thus losing one's freedom and peace. If something is acquired through honest means, there is no fear. If attained through dishonest means, one lives in perpetual fear. It is said that the one who never steals, God fulfills all needs.

4. Chastity

Sexual energy should not be wasted. It is a creative energy. Its waste depletes the body and mind, through tremendous loss of prana that could have been preserved and converted into higher energies. Its dissipation has the most depleting effect on both the psychic and nervous system.

Chastity is the strongest pillar in the spiritual edifice. The soul has no sex. Sexual thoughts or acts, except for procreation, degrade the soul. In married or unmarried life, chastity is essential.

With prayer and virtuous thoughts one can transform very little of the sexual energy into spiritual power. With chastity all the sexual energy is saved and stored up as spiritual energy, *Ojas*. It gives will power, intelligence and insight. One should endeavor to save one's spiritual energy by observing continence.

5. Non Covetousness & non acceptance of gifts

Non covetousness is a mental state devoid of cravings and greed, where peace and contentment prevail. Man's craving is the root of all unhappiness. Covetousness leads to longing. When unfulfilled it leads to anger, frustration and depression.

Practice of non covetousness brings the highest bliss of freedom. It exonerates us from the evil effects of fear, attachment, disappointment, anxiety, hatred, jealousy, anger and depression. It teaches us non-possessiveness, non-greed and unselfishness.

We have some basic needs to keep ourselves healthy and happy. We should find ways to simplify our life with fewer necessities to reduce time and energy spent in caring for them. According to research conducted by psychologists, there is no link between happiness and a larger income. In other words money cannot buy happiness.

Bondage is the worst bane. Acceptance of gifts invariably carries both bondage and indebtedness. Gifts very often hide an expectation of return from the giver. Bribery in guise of gifts or that will enchain someone to the giver is immoral. Whoever accepts such a gift loses his mental, moral and intellectual independence and becomes indebted and enslaved. It makes the heart and mind impure. One who keeps away from such gifts protects both his independence and integrity. His heart remains pure and his mind untarnished. Never accept a gift that implies bondage.

A gift is always imbibed with the negative traits of the giver. Whoever accepts such a gift will be left with those negative qualities. On no account should one accept gifts even it means hardship and suffering. Independence of thought, speech and action are our best gifts and should at all costs be preserved.

Gifts exchanged among close links, friends as token of love or keepsake are natural and decent. There is no harm provided there is no exaggeration or waste. Only gifts as token of love are not evil and do not carry any risk of degeneration or degradation.

Moral disciplines (Niyamas)

1. Purification

Purification means both internal and external cleanliness. Bodily cleanliness is external purification. Purifying the mind is internal purification. Although internal purification is of greater value, both internal and external purification are equally important.

External cleanliness is helpful to protect our health, to avoid infection and disease. Body cleanliness is carried out with soap and water. External purity without internal purity is no good. It encourages more attachment to the senses.

Internal cleanliness is mental cleanliness. A pure mind will not be tarnished by selfish motives nor polluted by negative traits. A pure mind will radiate unselfishness.

The repetition of mantras, sacred words or syllables and reading the scriptures or other sacred books, lead to internal purification. Practice of ethics is precisely meant for internal purification.

One should purify the mind by feelings of kindness and love towards all living creatures. Sharing and mixing with others as equals, goes a long way to establish peace with oneself, with our immediate links and with people at large. The circle keeps on widening, for the betterment of humanity. Peace and order will prevail.

Such thoughts and ideas that are painful and can do harm to others or are selfish and gratify our lower instincts should be erased. Our mind should be purged of these negative traits. Eliminate thoughts which are inimical, hostile, aggressive and obscure the mind. When a wave of anger arises, bring in a thought of compassion, forgiveness. If a dishonest thought comes, replace it by a thought of honesty. In this way all negative thoughts will be neutralized. The mind will become pure.

Cleanliness is the first step to overall progress and development. It does not befit anyone to be internally pure and externally dirty. One should observe hygiene laws, such as fresh air, pure water, washing of the mouth, cleaning teeth at least twice daily. Observe cleanliness of the body by a daily bath and change of clothes. Houses as well as surroundings should be kept clean.

Pure food keeps the mind and body pure. Avoid unclean, impure food which weaken our system and dulls our mind. Food sold with only profit making in mind, will make one impure, with the wicked thoughts. Food cooked for the sake of relieving hunger and prepared with all feelings of love will elevate one's thoughts.

Purity means cleanliness of both mind and body. Both external and internal purification have to be observed. Together they will yield better results. Purification of the body and mind will increase absorption of prana. The body works in perfect harmony with a higher buildup of prana. It brings us deeper concentration, an improved memory and sharp intellect. The result is perfect health and a happy disposition that can only promote peace and joy.

2. Contentment

Contentment is the one virtue that will thwart our greed, vanity and our craze to surpass others in possession, profession or influence. We should realise the ephemeral nature of phenomenal existence. Everything comes and goes. Their transitoriness does not enlighten our mind nor enrich our soul. Discontent robs us of our peace and serenity, leaving us obsessed or depressed.

Contentment is the state of total satisfaction, free from all desires or cravings. Irrespective of what we do not have or what we are experiencing in our daily life, we are ever happy. It is feeling sufficiency in whatever comes our way. Welcoming it as a gift of God, we are ever grateful that we have what millions do not or may not ever have.

Contentment is simplicity in living. It does not require much for simple living. Whatever our resources, we consider it more than enough. There is never any feeling of lack. The one immersed in contentment develops an inner feeling of joy and serenity, and experiences superlative happiness. There is neither envy nor desire of others' possessions.

The root of all misery are desires infesting and poisoning the mind. If unfulfilled, disappointment and depression is our lot. The eternal fever of desire gives us no rest, no peace: only dissatisfaction, distress, despondency and disease. The happiness we feel when we have forever destroyed desire far outweighs all the happiness of the universe.

3. Austerity

Austerity is discipline and restraint without denial. Avoid the two extremes of extravagance and asceticism. Everything should be done in the just measure without any denial. One should avoid excessive indulgence, be it in eating, sleeping, work or leisure.

We have to eat to keep healthy, so we have to be very particular about our diet. We should never overload our stomach, considering it a garbage bin. Occasional fasting will not harm anybody. It is a good way to give rest to the system and save vital energy.

One has the right to wholesome rest and enjoyment. We should never yield to any temptation which may result in bad habits or addictions. What we like is not always good for us. What we dislike may in fact be in our best interest. Rather than going according to likes and dislikes, we should bother more about what is best for us. With a little effort the senses will be under our control. The mind is the master of the body, not the body master of the mind.

Rather than just fulfilling our selfish ends, working for the good of others is also a form of austerity. It intensifies unselfish feelings in the human and urges him to selfless work. The heart is purified of its selfish motives. When we think of the good of others, we become oblivious of our own puny self. It is one way to curb our endless selfish desires.

The more one works to help others, the purer and less selfish one becomes. Working for the well being and success of all is the best way to rid ourselves of jealousy and envy. The feeling of oneness becomes a reality. Stress will then be nonexistent. Therein lies our peace, protection and ultimately betterment and perfect health.

4. Reading of spiritual books and scriptures.

Reading of spiritual literature will gradually restore our spiritual consciousness. It will help us progress morally and spiritually and become aware of the fleeting nature of physical life. Participating in religious activities, listening, reading or studying sacred scriptures fill the mind with pure thoughts. A pure, strong mind will revitalize our whole system. Only such books that will help us in our moral and spiritual progress should be chosen. We should avoid cheap literature. It is a waste of our precious time. It risks misleading us and hinder our progress.

5. Surrender to God

Surrender to the Lord is the highest expression of love and renunciation. Absolute, intense, unconditional love and supreme devotion is complete surrender to God. It is a deep and marvelous sense of dependence on the will of God. The individual will is surrendered to the Infinite will. Constant remembrance and yearning for God, with intense love and joy is also surrender.

One who surrenders completely to the Lord, always considers the Lord as the doer. He considers himself only an instrument of God to fulfill His purpose. He works free from any expectation, leaving the result to God. Love for love sake, worship for worship sake, without seeking, expecting or caring for results is complete surrender.

Work without any desire for result or reward. Work for the good of all, not for name or fame. Dedicate all effort and work as worship to the Lord. Whatever our activity our whole self should be focused on the Lord. Any action done should be taken as a blessed privilege from the Lord. Always work in humility, by sacrificing the ego on the altar of humbleness.

Giving is the centre, the basis, the gist of all moral teachings. Give generously without any expectation of return. Offer food and clothing to the needy, seeing God in all. Those who voluntarily bring knowledge and enlightenment to others should also be provided for their basic needs.

Anything done out of love gives it the highest meaning. It gradually culminates in complete adoration of God. We see God everywhere and in everything. Love of the divine leads us to universal love. Love all, hate none. Every act of love brings happiness. There is no greater happiness than to know or make others happy. Deify the world and everything in the world, for God and God alone permeates the whole of the universe.

CONCLUSION

The practice of truth and all the virtues purify the mind. The best way to practice these ethical principles is to replace negative thoughts by positive ones. Do good because it is good to do good! Ask no more. True morality is the path towards freedom. Immorality leads to bondage.

A pure and moral man has control over himself. That is why purity and morality have always been the object of ethics. Purity in thought, word or deed is among the highest ideals of ethics. A pure mind is man's most precious possession.

Ethics is the gateway to holistic living. Holistic living is the first step towards balanced health, peace and happiness which provide for quality living. The aim of all ethical laws is to eliminate those negative human traits which more than often are the cause of disease and suffering. Ethics bring forth man's finer qualities of compassion, love, unselfishness and intelligence. Ethics is the basis of a happy, healthy and harmonious life style.

THE EXPERIENCE OF GENUINE REST

YOGIC RELAXATION

The fast moving modern life has put all our everyday activities under pressure. Yet man is hardly aware of the growing tension overpowering him and sapping his energy.

Tension imposes on us a tremendous waste of vital energy, prana. Tasks that could be performed in all peace are overburdened with anxiety, resulting in severe depletion and loss of vitality. We live in a mighty competitive world which generates apprehension and anxiety through fear of failure. Many set the bar too high, inviting additional stress. Unknowingly, this ever increasing stress is having deleterious effects on the system. The body will try its best and within limits, either to adapt or to fight back. Failing which, it will collapse under the constant pressure of stress.

Our generation has witnessed tremendous advances in technology. It is a fast moving world and we are today just an incredible 'click' away from anyone, anywhere – in truth, from the rest of the world. On an individual level it is not easy to keep pace with this ever changing situation. The result is stress and tension. The remedy is relaxation.

Relaxation if practiced daily will go a long way to mitigate the disastrous effects of stress. That is why the seers of yore have made relaxation the basis of yoga. Yoga postures helps the body relax and rejuvenate. As a general rule yoga favours parasympathetic activation and diminishes sympathetic arousal.

Yoga includes many different relaxation postures: sitting, standing or lying down on front, back or sideways. Yoga relaxation is by far the simplest practice to teach and it is as easy to learn. These postures facilitate relaxation not only during yoga practice, but may be used to ease tension in different situations throughout the day. Deep relaxation on a daily basis will invigorate the system, strengthen its immunity and bring about other valuable physiological changes.

Stress can cause premature ageing of the body and the brain. Memory is impaired by recurrent heavy stress. Medical research has shown several health conditions that can appear or worsen under constant stress: asthma, depression, heart disease, high blood pressure, diabetes, obesity, allergy, gastric ulcers, colitis, irritable bowel syndrome, arthritis, reduced resistance to bacterial and viral infections and many others. Recent research has revealed that highly stressed persons are immunologically ten years older than their actual age.

Modern science has recognized that yogic relaxation techniques and yogic breathing are the greatest contribution of the ancient seers of India to mankind. Relaxation has been found to be effective in Alzheimer's disease, anger, hostility and aggressive behavior, anxiety, asthma, diabetes, fibromyalgia, headaches, hypertension and heart disease, insomnia, irritable bowel syndrome, panic disorder, substance abuse and smoking. Relaxation techniques have been used to treat side effects of cancer therapy.

Unless the frantic trend of modern life style slows down, tension and anxiety will stay with us. Psychosomatic illness or stress diseases, emotional pressure and wrong eating are some ills of modern living that medicine has to tackle. As long as this frenetic sway prevails we can only look for ways and means to alleviate these ills. Tension creates anxiety. Relaxation dissipates anxiety. Relaxation imparts to the whole organism a feeling of sublime peace and well being.

Yogic relaxation is not a state of inertia devoid of awareness. On the contrary one has a feeling of wellness in the body. One is very alert during the process at all times. Deep yogic breathing forms an integral part of yogic relaxation. There may also be repetition of a sacred syllable such as 'OM'. The whole system is harmonized with the increased inflow of prana. This will bring further serenity to the mind. The body is relieved of all stress and tension.

Many people have the bad habit of taking a nap during the day. It does not usually help. Instead of feeling relaxed one feels more fatigued. The body becomes sluggish. Yogic relaxation instead will bring all the benefits of increased vitality and a sense of rejuvenation. Muscles are relaxed. There is increased flexibility of the body. The complexion looks fresher. An improved appearance doubtless builds up a positive outlook.

Relaxation is an art and it is imperative that we learn this art to make ours that amazing feeling of rest and exquisite peace. Sitting, standing or lying down, one should let go a few minutes every now and then. It does pay dividends in terms of health and happiness. Whenever there is a surge of anger or increase in tension, there is nothing better to cool down than relaxation. Yogic relaxation prevents stress. The best way to control or prevent stress diseases is to adopt the daily practice of yogic relaxation. It is a must for health care irrespective of age.

Nothing compares with yogic relaxation where health and well being are concerned. It does not cost anything. One can be on one's own. No external help is needed. When the body is so relaxed, the mind experiences absolute calmness. When there is perfect coordination between mind and body there is heightened efficiency and stability in performance. There is also decreased health risk.

During relaxation the system is given a complete overhaul, leaving a feeling of unfathomable wellness and profound peace, never experienced before. The miracle formula for everyone lies in the daily practice of relaxation, whether health wise, for vitality or for a well balanced lifestyle.

CONCENTRATION – THE SECRET OF A SUCCESSFUL LIFESTYLE

Every human possesses the power of concentration to a greater or lesser degree. Knowingly or unknowingly we use this power in our daily life. In its simplest form it is the attention we have to pay to anything in our daily activities, reading, singing, cooking, etc.

Absentmindedness or lack of attention will make it impossible to assimilate knowledge or properly grasp the meaning. Very often, if our attention is not focused on any particular thing we do not notice it. Mental energy has to be completely focused to strengthen or increase mental power.

The power of attention is inherent in all beings. It is more a gift of nature, as in animals in their search for food etc. Geniuses in all fields, scientists, mathematicians, musicians, etc make full use of this power of attention. It is the source of all our knowledge and plays a vital role in our life. Through concentration we can better face the difficulties or emergencies of life and protect ourselves from calamities. Everyone would have fared far better in life had they been taught concentration from childhood. They are most successful who display great concentration. Indeed concentration is the secret of success in all spheres of life.

Dissipation is the root cause of all evils. Concentration is the essence of success in all spheres of life. Ninety nine percent of diseases, accidents or suffering is a result of our lack of attention to the laws which govern life and health in the world. The same attention to the observance of moral laws will bring understanding and tolerance towards fellow beings. Humanity will benefit by the resulting peace and harmony. The same attention if directed to spiritual matters will make us more virtuous and religious. It will help us unfold our spiritual nature. It will free man from the bondage of ignorance, delusion, selfishness and leads him to the attainment of health, happiness and harmony.

All the knowledge of the world has been gained through concentration. Mere theorizing has never produced any scientific truth. The tendency has always been to study and observe external things only. The vast majority is guided by the senses. Under the impulse of the senses the mind has the tendency to external attraction. Uncontrolled senses cause distraction. Control of the senses quietens the mind, leading to perfect concentration.

Most people use only two to ten percent of their potential. Only a small minority strives to make use of the power of the mind which represents man's unlimited potential. Every person should try to make full use of that wondrous power of the mind that is concentration. It is man's only means to knowledge.

RELAXATION, CONCENTRATION AND MEDITATION

It may be a good idea to dispel the misunderstanding commonly met with, concerning these three different techniques.

Relaxation should not be confused with concentration, nor concentration with meditation. It is true that relaxation does help concentration and concentration helps meditation. Each is a different practice, yielding different results.

RELAXATION

Relaxation eases both physical and mental tension. It helps one feel comfortable in one's body, a complete let go. Tense muscles become relaxed or loosened. One gets rid of fatigue. There is not the least tension left in the body. It will gradually lead to a diminution of mental tension. Although there is awareness of both body and mind, there is hardly any physical or mental activity. To help ease tension one may concentrate on something or other. One may follow the breath, inhalation or exhalation and synchronise them with the repetition of a sacred syllable like 'OM'. Rhythmical yogic breathing will further reduce both physical and mental tension.

CONCENTRATION

Concentration is the mental faculty of focusing the attention on some specific thought, object or event. In daily life, circumstances sometime compel us to think deeply, especially faced with a problem. The mind focuses on the problem and we are lost to the outside world. Concentration deepens. The body is forgotten, the senses are withdrawn. The mind is lost somewhere. We are wide awake, yet we no longer feel our physical presence.

The mental energy brought to focus on one point is known as concentration. Unbroken concentration or perfect concentration leads to meditation.

MEDITATION

Meditation is a higher form of concentration. It is a spiritual practice of a superconscious order. Meditation involves yogic breathing and repetition of some sacred word. The body becomes motionless, yet there is awareness of its existence. The mind is absolutely thought free and it is witnessing a higher state.

The higher the concentration the higher the mind will reach into the realm of higher consciousness, Superconsciousness. This is meditation. By continued meditation the subtler, superconscious state is achieved. The conscious mind becomes merged into the superconscious, revealing a bright inner vision.

Coming out of concentration, one is the same old self. After coming out of meditation, one's perception is altered, having discovered a higher state of existence.

MEDITATION – DISCOVERING OUR TRUE IDENTITY

In meditation there is perfect concentration of the highest order. The mind is not disturbed by external or internal factors, by any exterior noise or by any disagreeable thoughts.

Meditation is the only means to check or control the onrush of the mind. When the mind is under control, it becomes calm. Perfect calmness or balance is the real nature of the mind. Pure consciousness fills the mind as the pure thoughts are transformed into pure consciousness. This will facilitate meditation and help to enter the state of superconsciousness.

The mind has three states, the subconscious, the conscious and the superconscious. The highest is the superconscious. From the conscious state, during meditation one can experience a higher level of consciousness, the superconscious state of the mind. When the mind has attained to that superconscious state it is known as Samadhi, the state of perfect meditation. Samadhi is absolute peace and absolute bliss. To realise this state is to discover the real value of life.

The conscious state is limited to gathering knowledge which will be transferred and stored up in the subconscious. This stored knowledge will unconsciously influence both body and mind. The subconscious state manifests as instinct, habit, memory and acquired knowledge.

In meditation there is barely any activity as the body becomes motionless. It requires a minimum of energy. The valuable energy thus saved will be transformed into higher consciousness. The breath slows down to a faint motion hardly perceptible or audible. As meditation deepens there is hardly any sensation of the body, only the awareness of its existence!

Meditation brings detachment from the senses. Yogic breathing help prevent the mind from getting distracted by the senses. The mind is turned inwards to concentrate upon itself. In meditation, the respiratory rhythm changes. It gradually slows down to become barely perceptible. In deeper meditation there is an exquisite floating sensation. One can experience sublime inner visions and a wonderful sensation of weightlessness.

Prana and mind are closely linked. Thoughts are the finest vibrations of prana. Once prana is controlled and rhythmic, the intense activity of the mind will calm down. It will help the mind go higher and higher to attain the superconscious state. Until the mind is perfectly controlled, balanced and silenced, it is difficult to reach the state of meditation.

Only when concentration is very deep can one enter the state of meditation. As meditation advances and becomes deeper and deeper the state of superconsciousness is attained. The mind is transformed into pure consciousness. Superlative peace permeates the core of our being. We have come face to face with the highest plane of consciousness, beyond physical existence. This is the objective and ultimate goal of the science of meditation.

There are three stages in meditation. In the first stage, Pratyahara, the mind is withdrawn from the senses. In the second stage, Dharana, concentration proper, the mind remains in focus on some object, indefinitely. There is not the least distraction. Repetition of a sacred syllable like 'OM' will help rid the mind of the hundred and one thoughts that constantly plague it.

Concentration is not meditation. Perfect concentration, Dharana, will slowly lead to meditation, Dhyana. Meditation reveals to man his own soul. In meditation the mind is turned inwards upon itself. Man's consciousness expands. The more he meditates the more and more he will expand beyond consciousness into superconsciousness.

The science of meditation has as its basis the science of yoga. Bhakti yoga, Karma yoga, Jnana yoga, Raja yoga, as all other yogas are so many ongoing stages in preparation for meditation. They have one and same objective, the awakening of the Kundalini lying dormant in every human. The awakened Kundalini makes meditation easier. The early stages of most yoga disciplines, ethics, like truthfulness, non violence and all other virtues are practiced to quieten the mind in preparation for meditation.

Siddha kundalini mahayoga is different from other yoga techniques. Here, the awakening of the Kundalini depends entirely on the guru's grace. The siddha guru transmits his divine powers into the recipient to awaken the dormant Kundalini. Meditation experience will occur spontaneously from then onwards. However, to progress in meditation the recipient has to master a sitting posture and practice yogic breathing as prescribed in all yoga disciplines. Ethics too are fundamental. The recipient has to observe the restraints and disciplines, the Yamas and Niyamas.

The aim of restraints and disciplines is to purify the mind. The finer and higher the thoughts, the higher the plane the mind can exist. The ethics are as fundamental to spiritual progress as in our everyday life. No one is an exception. Both mind and body have to be pure for one to progress spiritually. The ancient seers of India stressed the importance of both internal and external purification. These help convert the physical and mental prana into higher spiritual energies, Ojas, to facilitate meditation. If not wantonly wasted, all the forces that are working in the body can be transformed into Ojas.

Repetition of a sacred word such as 'OM' is indispensible for access to the great source of spiritual power. These sacred words, together with yogic breathing and a thoroughly mastered yoga sitting posture form the basis of the science of meditation. Yogic breathing is the easiest way of getting control of the vital energy, prana, in the body. Proper diet is also important. All the forces working in the body come from our intake of food and the prana therein. A poor diet weakens the body and the mind.

Meditation gives intensity to our thoughts. Even scientists unknowingly use this power of meditation for all their startling scientific discoveries. That is the secret of their success. They are lost in their subject, forgetting their own self and everything else. They are cut off, although momentarily, from their physical body. Only their search exists in their mind until the truth flashes within.

The average man is assailed with worries, anxieties, fears, blunders and endless tribulation because he is only body conscious. As long as body identification remains, man will remain a slave to the senses. Even his sleep is burdened with dreams and nightmares. Meditation shows the way out by turning the mind within. Only meditation brings genuine peace and happiness. This is the ideal way to unwind the entire system. One or two hours of meditation are worth the most marvellous and superb rest. A rest never experienced before. It is joy sublime!

The whole science of yoga and meditation is to help man understand his true identity. He is beyond body, he is beyond mind, he is Spirit.

KUNDALINI – THE POTENTIAL POWER OF MAN

Kundalini is the primal energy, the divine cosmic energy, the potential power of man. Life depends on this dormant energy for its transformation, dynamisation and sublimation. It controls man's physiological, mental and spiritual evolution. As divine energy it is latent in man and needs to be awakened. Unless it is awakened, man will not have any spiritual growth. It is the basis of all spiritual experiences. Its awakening makes man spiritually strong.

As physical energy, it is very much awake and alive. It is behind man's physiological growth, his every action and thought, word and senses. Along with prana, it pervades the body and keeps man alive. On the physical level, it controls his very life breath, his heart beat, circulation, respiration etc. It is the basis of life. As the life sustainer, Kundalini is very much active. It operates through the three lowest centres to provide for physical needs only: Muladhara centre where all the physical energy and spiritual energy are concentrated; Svadisthana centre and Manipura centre, where considerable physical energy and prana are stored.

Kundalini is the symbol of man's divinity. It is present in every human being, although it remains dormant in the majority of cases. The average person is little removed from his lower instinct. Eating, drinking, sleeping and procreating are common to all creatures. It is only the awakening of the Kundalini that distinguishes man from animals. Its awakening marks the beginning of his great spiritual journey. The higher man rises spiritually the higher he is above anyone else.

The word 'Kundalini' still evokes mystery. It is as fascinating and as awe-inspiring. This Divine gift brings as much of a thrill for those eager to discover its presence. For them it is the most delightful and wondrous experience. To those who hold it in utter awe, it gives a downright cold chill. For them it is most dangerous.

Kundalini is the unique way to attaining superconscious perception, realisation of the spirit. It is the secret to divine wisdom. Sages of all times and climes discovered this inner power to man's perfection. The Kundalini once awakened gradually brings the mind and senses under control. These Sages of distant past understood the mind as the mightiest instrument of knowledge. Control of the mind alone will unveil the highest knowledge of the self.

Man's mind, utterly externalized, has lost all notion of the subtle world within. In normal life everything man imagines, sees or dreams he perceives in the external world. When the Kundalini is awakened, perception is also in the mental space. It is the sublime experience of the ancient seers. For the first time, within the once dark abyss of his mind, man discovers the ineffable realm of consciousness. He realises the mighty difference between ordinary consciousness and superconsciousness. Then alone is the mystery of life revealed.

Consciously or unconsciously all prayers, worship, rites, rituals and ceremonials ultimately lead to the awakening of the Kundalini. All good thoughts, actions, prayers, an exemplary life, a strong faith and longing for God transform a part of the physical and sexual energy into spiritual energy and give man more and more spiritual power. Through longing and devotion to God this infinite inner power, coiled up in sleep in the Muladhara centre, will be awakened to push its way through to the Sushumna, located in the spinal cord.

The goal of all spiritual practices is to awaken and cause the Kundalini to ascend from the lower centres along the Sushumna, Muladhara, Svadisthana and Manipura, to the higher, Anahata, Vishuddha and Ajna and the highest, Sahasrara. To raise the Kundalini is the whole aim of all yoga practices. Therein lies all knowledge. It can be awakened through mantra-repetition, mantra yoga; through yogic postures, breathing, bandhas and mudras, hatha yoga; through philosophy and knowledge, Jnana yoga etc.

Kundalini is the only means to succeed in meditation. It lies dormant and will only manifest as a spiritual force, when awakened. Its awakening is a natural process. It is in no way dangerous especially if awakened through Divine grace or by the grace of a guru through shaktipat. In shaktipat the guru transmits his own shakti, divine energy, into the recipient. The awakening is spontaneous and by far the simplest and easiest way.

With the awakening of the Kundalini, the recipient will immediately or within a short time experience a higher plane of consciousness. The dark recesses of his mind will glow with the luminous form of consciousness. While the experience lasts, the mind will be still, as his consciousness has ascended to a higher plane. From outside he may give the impression of being motionless as in sleep. However, he has perfect awareness of being merged into a higher plane of consciousness. He becomes witness of his own consciousness. It is a state of blissful being.

While the awakening is possible through a guru's grace, its raising depends entirely on the recipient. Once awakened the Kundalini does not go back to its dormant state. Anytime it can be made to rise, but only through sincere personal effort and a disciplined life. One has to follow the divine rules and put in the right effort. Otherwise the Kundalini will not ascend. One will not progress. Yogic breathing is important, as mental repetition of a sacred word received from the guru. Living according to ethics is fundamental, as proper diet. Only then can one slowly raise it from centre to centre till the brain centre, Sahasrara is reached.

Every being that has a spinal cord has the three psychic currents present in them. The Sushumna is the main of the three vital psychic currents. It appears as a luminous live thread within the spinal cord. Ida and Pingala are the two lesser psychic currents. Vital force, prana is stored up in the sushumna. It is indeed the live passage of man's salvation. Its extremities end in two centres, Muladhara at the base of the spine and Sahasrara, at the apex. Between these two are five other subtle centres. These seven centres are so many planes, each corresponding to a higher layer of the mind.

The latent energy has to be awakened and directed through the sushumna up the spine to the brain. As the Kundalini moves from one centre to another a new superconscious world unfolds within. Its ascent is more of an unfoldment, expansion and elevation of consciousness until it attains its blissful state in the brain centre, Sahasrara. Once it reaches the Sahasrara it is the highest state of superconsciousness, Samadhi. Sahasrara is the ocean of Infinite Light, Knowledge, Existence, and Bliss.

YOGA – THE HOLISTIC METHOD PAR EXCELLENCE

HATHA YOGA

Yoga is a complete science and art. It is a unique system the world over. Yoga does not highlight physical health only, but also ensures mental and spiritual health. Yoga comprises a series of holistic practices involving self discipline, meant to purify and bring into close harmony body and mind. It works to perfect the body, filling it with life force, prana. It helps repair or compensate for physical defects. It allows gradual realisation of the highest human potential.

From east to west, north to south, everywhere these extraordinary yoga postures and philosophy have conquered millions of hearts. They are so popular that they have become an everyday household word. From the simple housewife to the most experienced medical scientist, all are agreed that yoga has an important place in the maintenance of physical wellbeing.

Yoga is highly beneficial in the management of psychosomatic disorders. It has proved itself as the ideal remedy to stress and stress related diseases. It helps counteract the effects of inactivity, thus decreasing depression and anxiety. It is nowadays being recommended by modern science for patients having such disorders. There is sufficient proof of its positive role on the psychosomatic frame. In brief, yoga is of the utmost benefit to the whole organism.

Human health depends on the excellence of the cells and tissues. This in turn depends on proper food and adequate absorption of prana. Yoga practice encourages a balanced diet and regular eating habits that is psychologically beneficial and health promoting. Correct breathing is indispensible for the intake of prana and oxygen the organism requires for good functioning. Ignorant of its importance or what is at stake, the majority of people do not know they have to pay particular attention to their breathing. Yogic breathing and control of prana come as a blessing to those afflicted by disease, as to those who want to protect or maintain their health.

The body has the amazing capacity to heal itself in optimum conditions. This is enhanced by yogic breathing and postures which increase the supply of oxygen and prana to the body. Correct breathing is critical to human health. Hatha Yoga teaches various breathing exercises, the most important being deep yogic breathing. Yoga increases strength and flexibility, improves circulation and promotes well-being. It acts as a preventive agent to the common ills. It helps preserve health and increases mental alertness and longevity.

Unlike modern medicine which is largely a science of disease and treatment, yoga is a science of health. According to yoga, most diseases, mental, psychosomatic and physical originate in the mind through wrong way of thinking, living and eating. The basic approach of yoga is to correct a negative life style by cultivating a rational positive attitude. Performance of the postures necessitates a calm mind. Having to fight mental unrest will invariably waste precious prana. One of the reasons, why the practice of restraints and disciplines are fundamental to any yoga practice.

Hatha Yoga should not be confused with simple physical exercise, although it deals mainly with physical postures. Perhaps the one path of yoga the most widely practiced, yet it is the most misunderstood of all. Unlike other physical exercises, the system of hatha yoga is a combination of postures, deep yogic breathing, bandhas, control of prana and concentration. Yoga postures are not for body building although they promote health. Benefits from yoga postures are much more compared to any other system of exercise. Yoga is finest blend of physical, mental and spiritual exercise.

Today it is universally recognized by layman and scientist alike, that yoga has a crucial role to play in the preservation of the vital energy for protection and promotion of health. Regular yoga practice ensures a smooth and prompt return to homeostasis following experience of any imbalance, internal or external. It is probably the most practical method to induce health, vigour and a better and active life style.

Yoga also prepares one for a spiritual life. It reveals the rich inner life of the spirit, the power of clear and better thought. It incorporates the fundamentals and completeness of a spiritual practice. The discriminative faculty is strengthened. One is a changed person, for the better. There is poise, confidence, self discipline and a greater incentive to free oneself of bad habits. The postures, breathing and mental repetition of sacred words accentuate the absorption of prana which activates the body at all levels, mental, physical and spiritual.

That is why hatha yoga forms the preliminary practice of various yoga disciplines. Of divine origin, many yoga postures, like the lotus posture, are meant essentially for the purpose of sitting in traditional prayers and especially during long hours of meditation.

Yoga postures, breathing and meditation helps one gain physical strength and flexibility as well as a calm mind. Yoga and meditation also promote the spiritual unfoldment of the individual. In truth, the spiritual dimension of yoga is the very basis of any lasting improvements which may be noted at all levels.

With growing faith, there is greater emotional ability to eliminate nervous tension and improved mental poise. One breathes life. The mind/body/spirit balance is restored. The person has more awareness of his being. He is not just a mere body, but mind and spirit also. All three have to be in optimum condition for him to be healthy. Much depends on his own will power, on his own willingness and sincerity in integrating the three facets of his life to emerge as a complete being.

YOGA POSTURE BENEFITS

Hatha yoga is the science of health, incomparable in many ways. Yoga postures have beneficial effects on organs of digestion, elimination and on the circulation. They give the organs an internal massage. Muscles undergo considerable, gentle stretching and contracting and help maintain elasticity.

There is no brisk movement in the performance of postures. They should be done slowly, with grace and rhythmically. Gradually all stiffness of limbs and joints disappear. Postures require holding of specific areas of the body followed by relaxation, along with deep yogic breathing. Postures stimulate key pressure points that govern the flow of vital energy, prana.

Holding or maintaining the posture is the most important component of yoga performance. Maintenance of posture will allow the prana to work that part of the body involved. In postures where there are several stages, each stage has to be maintained for one or two breathing rounds for maximum benefits. Mental repetition brings calmness to the mind and more concentration in the performance and memorising the posture.

While performing an asana, the body is brought in stages, in a certain position. Each stage has equal importance as the final posture. The same attention has to be given to the performance of each single stage. At each stage a specific organ system is involved. Prana is directed to that part of the body to relieve pain and heal disease and restore its balance. There is an internal massage of the different organs involved in all different stages of postures. This stimulates, tones and rejuvenates all the muscles and organs.

Yoga stretches not only the muscles but also the soft tissues of the body, the ligaments, tendons, and the fascia sheath that surrounds the muscles. There is sustained backwards and forwards flexing and stretching of the body. Increased lubrication of the joints, ligaments and tendons bring increasing flexibility. This results in a sense of ease and fluidity all over the body.

Hatha yoga will be more beneficial if practiced along with deep breathing and a calm and concentrated mind. To calm the mind, mental repetition of a sacred syllable is ideal. Some sacred word as 'OM' or any other sacred word should synchronise the breathing. As the mental repetition and the breath become perfectly synchronised and rhythmic, the whole body slowly and gradually become rhythmical, bringing perfect relaxation to the whole body. Wherever possible, closing the eyes may further help concentration.

Most postures are beneficial to the circulatory system. Abdominal lift, although needing time to be mastered is an incomparable and wondrous exercise. It takes care of the heart, the main organ of circulation through the sustained movements of the diaphragm. Headstand, shoulder stand and inverse postures, all being inverted postures, help the blood to flow more easily back to the heart.

Most veins have to work counter to gravity to collect blood from all parts of the body and return it to the heart. This is a difficult uphill course. Because of the nature of their work and many through ignorance, keep standing for long hours. They may suffer from varicose veins because of this added strain on the venous system. Practiced often, inverted postures minimize the pressure on the veins. The veins are amazingly rejuvenated.

Head stand and inverted postures allow for a richer supply of blood to the brain. They promote the mental faculty, memory, sharpness, self control, determination and discrimination. The brain requires three times more oxygen than the rest of the body. Thus, increasing oxygen flow increases alertness and encourages better mental functioning. They retard ageing processes and prevent such diseases as Alzheimer etc.

Locust postures are excellent for the lungs, thoracic region and the muscles of the back. Cobra, bow and locust postures are excellent for stretching the abdominal region and contracting the back region. Forward bend, stork, plough postures contract the abdominal muscles while stretching the back. Twist postures will exercise the lateral muscles of the abdominal region and give added flexibility to the spine. The waist benefits from such postures as cobra, bow and abdominal lift. Shoulder stand, inverse, fish, lion postures are marvellous for the prevention of throat problems and common colds.

The aim of yoga is to bring equilibrium to the whole system. It would not be advisable to concentrate on just those exercises mentioned for specific reasons. This approach might itself cause further imbalance in the body. The ideal would be to perform a daily routine comprising of a series of postures that will exercise the whole body systems. Our daily series have been designed in keeping with this approach.

Corpse Posture – Shavasana

One of the basic yoga postures, shavasana provides complete relaxation. It is the ultimate relaxation pose, the most perfect posture for rest. Every system gets completely relaxed. In shavasana there is practically no energy or prana being used. Only a minimum is used to keep the body going, the remaining being stored as reserve. So long as the body is in a state of tension, prana and other subtle forces cannot work in harmony.

When deep yogic breathing is done in shavasana the flow of prana becomes rhythmic with even distribution throughout the system. There is a general decrease in metabolic rate and oxygen consumption. The nervous system gets completely rested. The arteries and veins are under less pressure. It will result in a lowering of blood pressure levels. The resulting decrease in heart rate will relieve the heart of undue activity.

Shavasana relieves us of tension. Muscle tension is reduced, leading to decreased stress levels. By relaxing deeply all the muscles one can thoroughly rejuvenate the nervous system and attain a deep sense of inner peace.

This relaxed feeling is carried over into all one's activities and helps us conserve energy and let go of all worries and fears. There is a marked decrease in fatigue, coupled with deeper and sounder sleep. It will lead to improvement in concentration and in memory. Our daily performance will improve.

When the body and the mind are under constant pressure, their natural efficiency diminishes. Modern life style makes it difficult to relax. We have forgotten that rest and relaxation are nature's way of recharging. By releasing the tension in the muscles and putting the body at rest, the whole system is revitalized.

Very often under heavy stress or fatigue, our brain blacks out for a moment. We are left staring blank in space, motionless. We are so cut off that no external disturbance, noise or presence affects us. We are oblivious of our surroundings. It is an involuntary, 'negative relaxation' imposed by extreme fatigue. Such unexpected incidents may be disastrous.

To avoid these extremes we should give our system the break it desperately needs at times. Let it become a daily feature that we consciously go into corpse posture relaxation for a good five minutes. This should be done at least once or twice, not to accumulate tension, stress or fatigue. It will spare us being constantly on the verge of collapse.

Everyone should find the time daily to give the body some rest. Even amidst the bustle of daily life there is always a way to steal a break between these endless responsibilities and allow the body the little rest it so badly needs. These 10-15 minutes in shavasana are worth more than a full night's agitated sleep.

SALUTATION TO THE SUN

SURYA NAMASKAR

Surya Namaskar, salutation to the sun is of ancient Vedic origin. *'Surya'* means sun, *'Namaskar'* means salutation. It was a spiritual practice, paying respect to the sun god; a form of worship handed down from ancient seers.

The sun was daily worshipped as the most powerful symbol of spiritual consciousness and the source of life. Adoration and worship of the sun was probably man's first attempt to some form of higher expression.

The sun is the most powerful and concrete form of the almighty God. Until today sun worship is practiced in India. It is believed that those who worship the sun as the creator lord become powerful, intelligent and enjoy longevity. The Egyptian, Aztec, Inca and Mayan civilizations also worshipped the sun.

The sun is the source of all life on earth. Surya namaskar is direct worship to the sun, the prime provider of prana, vital energy. It is always performed facing the sun, with deep yogic breathing and the repetition of sacred words. It enables us recharge our physical and mental energies. This will help maintain good health, mental calm, a strong body with increased stamina, physical and mental. A healthy body and a composed mind are the basis for spiritual unfoldment.

Surya namaskar with its multiple combinations of movements thoroughly exercises the whole body. The poses are arranged so that they alternate flexing and stretching of the spine and body. It intensifies the natural flow of prana through the whole system and improves body functions. Once mastered it is effortless and delightful to perform. It does not require much time either.

Surya namaskar may be considered a complete practice by itself. It is ideal for the very busy with little time to spare. There are fourteen postures in surya namaskar, performed in the following sequence: Prayer pose – *Pranamasana*, Palm tree posture-*Talasana*, Crescent-*Ardha chandrasana*, Stork-*Padhastasana*, Equestrian posture-*Ashwa sanchalanasana* , Crescent lunge posture-*Anjanaya asana*, Elephant-*Gajasana*, Log-*Chaturanga dandasana*, Cobra-*Bhujangasana*, Cat-*Bitilasana*, Sun salutation-*Ashtanga namaskar*, Thunderbolt- *Vajrasana*, Swan-*Hansana* and Squat-*Malasana*. A full round of surya namaskar is considered to be of two sets. In one set, there is a sequence of forty postures. A complete surya namaskar will thus comprise 80 postures!

In the first set we start with the right leg stretched back. For the second set the same postures are repeated, this time starting with the left leg stretched back. Ideally these should be performed in a continuous, unbroken flow. If the strain is too much, complete the first half; relax in corpse posture before doing the second half. It takes 8-10 minutes to complete one full round.

Surya namaskar gives tremendous stamina and strength to the whole body that no other form of exercise can claim. Complete in itself, it can conveniently replace a day's practice when short of time.

It is not possible for a beginner to attempt the complete surya namaskar at one go. One should become familiar with the different postures of the series before attempting the complete set. Each posture has to be performed until individually mastered. The body will welcome the gradual change and respond better to the new movements. Only then will one be able to perform and understand the import and beauty of this unique combination.

There is no harm to break the sequence until all movements are mastered. Do not rush through the postures. Do not switch abruptly from one position to another. Take your time and perform slowly and gracefully, with maximum stretch to the body. Hold each posture a while before changing to the next.

The next step is synchronizing the breath with the movements. The breathing sequence synchronises with each posture and makes its performance easier. Extension is always accompanied by inhalation, flexion by exhalation. Exhale during contraction. Inhale during expansion. This will avoid strain and waste of energy.

Surya namaskar is mostly performed with only the hands and feet in contact with the floor, the rest of the body being raised. Only for ashtanga namaskar, eight points of the body: the feet, knees, palms, chin and chest; and for swan, seven points, are in contact with the floor. However there is no harm for beginners to rest other parts of the body on the floor while learning.

Surya namaskar may be practiced independently at sunrise. Postures should be performed with the minimum of effort. The movements should flow smoothly after each other, to make it more beneficial and enjoyable. One complete round will do for beginners. Those who can afford to practice more in terms of time and endurance may perform two to three rounds.

YOGA COURSE

This yoga course has been structured with a very scientific approach. One goes gradually, in easy steps from the very simple postures to the more advanced ones. The body has the time to adapt and progressively gain suppleness and elasticity. It is thus able to perform the new postures and movements with better ease.

The course is in many ways unique. Some 108 postures are arranged in weekly series of different postures. The first eighteen weeks are devoted to learning new postures weekly and reviewing those that have been previously learnt. The step by step illustrations will facilitate the practice. The lessons are arranged so that the sequence of movements is smooth, changing gradually from simple to more complex. This will help increase suppleness and flexibility. Upon completion, the postures are rearranged into daily series, Monday to Saturday, to be repeated every four weeks. Complete in itself, the course is suitable for a lifetime practice.

The remarkable results we have obtained through yoga and meditation in almost four decades of practice are quite impressive, simply astounding. Sometimes, even doctors would recommend our institution to such patients where they remained helpless. Many professionals from the medical field too, came for relief. The volume of healing cases has been astounding especially the many cases given up medically as hopeless. The cases include not only psychosomatic conditions, but also serious physical and mental pathologies.

Almost immediately after starting meditation and yoga, all noticed changes in their condition. There was marked decrease in intensity and seriousness of their disease condition, significant alleviation of symptoms, sometimes complete remission or cure of the disease. Most patients would gradually reduce their medication within one to four weeks. There were even a few cases where all medication could be discontinued overnight.

However, we always advise patients to continue their basic medical treatment regimen until their condition is cured or controlled. In chronic cases like diabetes mellitus, epilepsy etc, it is advisable to continue with medication, albeit in much reduced doses to prevent any relapse or complications.

It is true that practice of hatha yoga, especially for those who are sedentary, will make them lose many extra kilos in the beginning. But this loss in weight will slow down as they continue the practice. One should not expect results too fast. The yogic method proceeds in a slow systematic way and does not try to radically alter anything at one go.

Beginners may experience some stiffness in their body but it should not discourage anyone. Always start the day's program with a few warming up postures. Allow time for the body to 'wake up'. Muscular tension or excess body flab will gradually disappear. Tightness in tendons and ligaments will be eased. Joint stiffness will lessen, making them more flexible. There will be more suppleness in the whole body. With practice one feels a sensation of lightness. The body is completely relaxed and rejuvenated.

GUIDELINES FOR YOGA PRACTICE

1. Always face East or North during practice. When lying down, head towards East or North.
   (Exception: For Surya Namaskar, follow the direction of the sun till sunset. Otherwise, face east or north.)
2. Practice on an empty stomach. If not feasible, a glass of plain water, juice or milk, a few nuts or fruit in very small amount, a light snack such as yogurt, may be taken 30 minutes before practice.
3. After practice, allow 30 minutes before taking a light meal.
4. Allow 1 hour after practice before taking a full meal.
5. Allow 3 hours after a heavy meal before practice.
6. There are no dietary restrictions, although in the long run it may enhance the practice.
7. Ease bladder and move bowels before practice. Relieve yourself if necessary during practice.
8. Wear light clothes. Avoid tight outfits, tight waist bands, buttons and zippers. Elastic waist bands are best.
9. It is best to practice 2 hours before sunrise, between 4-6 a.m or after sunset, 6-8p.m. Otherwise, any convenient time will do. Go to sleep latest by 10 p.m. Get up 2 hours before sunrise.
10. Use a blanket, a thick folded bed sheet, or a foam 1-1 ½ inches thick, a rug or carpet.
11. Practice in the open or in a well ventilated room. Always keep windows open.
12. Try to practice at one permanent spot. It helps create a positive atmosphere.
13. In the beginning, practice some half an hour. One may increase gradually after completion of whole course. Increase the number of breathing rounds for each posture or do some extra postures from other days.
14. A shower or bath is not necessary before practice.

15. Allow 30 minutes after practice for a bath or shower. Avoid too cold or too hot water. Luke-warm bath is best.
16. Go into the postures slowly. Unlike other physical exercises which involve quick movements, yoga involves slow, easy and graceful movements.
17. Do not strain! Perform according to your capacity, with ease and grace.
18. Relax for 3-5 breathing rounds between postures.
19. After completion of any yoga session, relax in corpse posture for at least 5-10 minutes.
20. Anyone above the age of puberty, 12-14 yrs, can practice yoga. Children under 12 should avoid, unless in special circumstances, under strict supervision. However, they may practice yogic breathing. From a tender age all children should be taught low, abdominal breathing, through the nostrils. This will help them keep healthy and avoid all illnesses linked with improper breathing.
21. Women should avoid all heavy postures during their menses.
22. Women should not start yoga practice during pregnancy and should wait at least 3-4 months after childbirth to start.
23. Pregnant women already practicing yoga may carry on up to the 7th month. After childbirth they may resume yoga after 2-3 months.
24. After practice we should feel quite fresh and in shape, both physically and mentally. Improper practice will cause fatigue. We should try to reassess our practice and find any fault.
25. In case of illness, seek medical advice first.
26. Any form of addiction should be tackled gradually, not to cause any imbalance.
27. PRACTICE REGULARLY! The beneficial effects will usually come after a few weeks. There are many cases where beneficial effects are felt within days of practice. Others, within one or two weeks.

WEEKLY YOGA PROGRAM

Some postures have several Sanskrit names. We have as far as possible given the one most widely accepted *(Italics)*.

FIRST WEEK

RELAXATION POSTURES

Yogic relaxation involves several relaxation postures. The corpse posture is one of the most important postures. It gives absolute relaxation. Its effect is immediate and amazing. Body and mind both experience perfect calmness. Done 5 to 15 minutes, it is worth more than several hours of unsound, restless sleep.

The highly strung person should find some few minutes to relax. Therein lies the secret of vitality and peace for those with a hectic life style. For the first time perhaps, we will discover what rest or relaxation really means. We will find mostly lying down postures, the best way to start.

It is much easier to relieve one's stress and anxiety through relaxation rather than any drugs or medication. Some 5-10% of people around the world are taking sleeping pills. If we have trouble sleeping at night these postures will surely help us fall asleep without looking for any external aids. Recent research has shown that sleeping pills can give rise to an increase in cases of early deaths and even some cancers.

In the state of perfect relaxation our body becomes limp and motionless. It is really a world of difference – another world, tension free with a more positive outlook, more pleasant, full of hope! Even beginners tend to feel less stressed and more relaxed after the first lesson.

No one has to learn how to lie down, but most of us may have to learn how

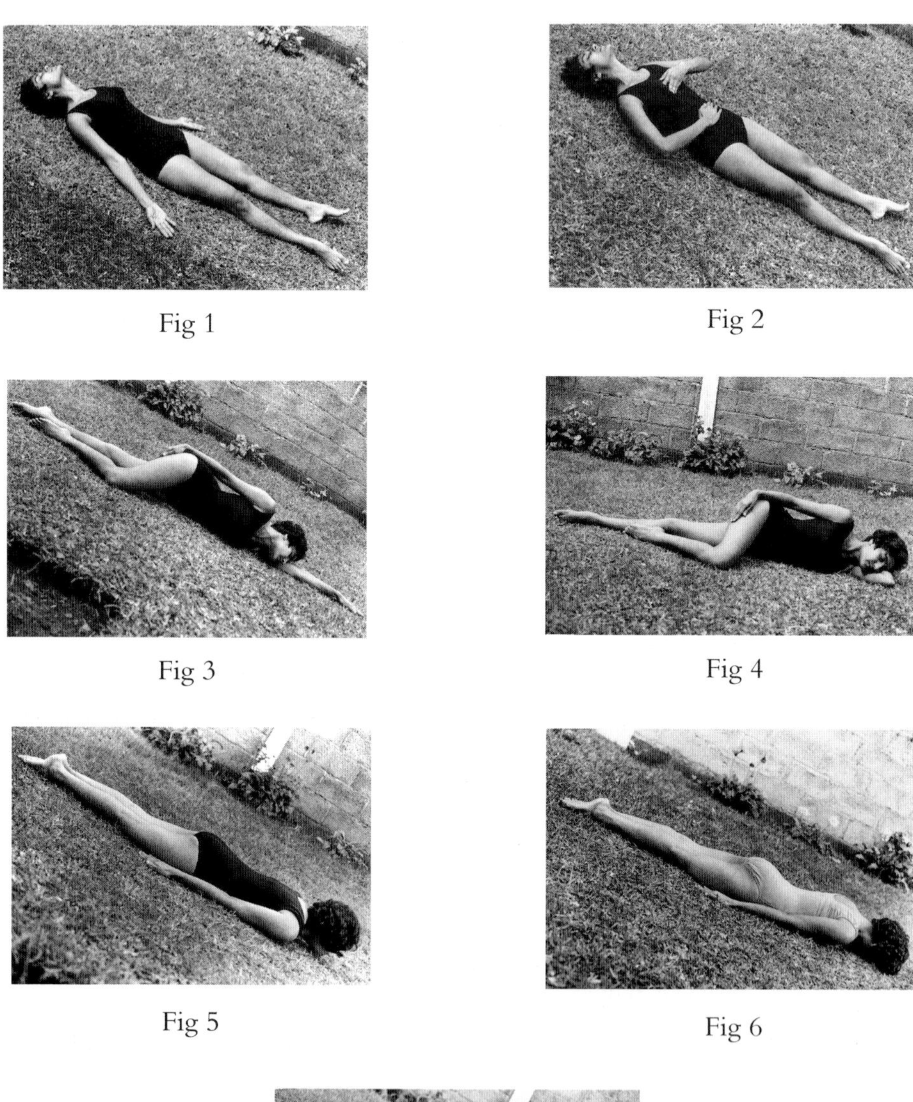

Fig 1

Fig 2

Fig 3

Fig 4

Fig 5

Fig 6

Fig 7

to breathe properly. Breathing has to be synchronised with repetition of some sacred word. It helps calm the mind and prevent distraction. In Hindu tradition, the sacred word 'OM' is the most widely used.

Corpse Posture – *Shavasana*

1. Lie flat on back, head resting on crown, chin up.
2. Keep arms away from body, palms up; legs apart, toes pointing outwards. (Fig 1)
3. Relax completely for 5 minutes, doing deep breathing as below.

Deep diaphragmatic breathing – *Deergha Swasam*

1. Lie in Corpse Posture.
2. Place one hand on chest and the other on abdomen. Breathe out. (Fig 2)
3. Breathe in slowly and deeply through nostrils. Breathe out slowly. This makes one breathing round.
   Breathe to the ratio 1:2. Breathing in 1: Breathing out 2.
   Remember, only the hand on the abdomen should move up and down. The other hand remains motionless.
4. Complete 3-5 breathing rounds.

Lying On Side – *Dhridasana*

1. Lie on left side with left arm either stretched (Fig 3) or bent (Fig 4) and serving as pillow for head.
2. Bring right leg over left leg, the right knee slightly bent. Rest right arm loosely on right hip. Breathe out.
3. Complete 3-5 breathing rounds.

Lying flat on tummy – *Udara Shavasana*

1. Lie flat on tummy, arms alongside body, palms down. Forehead (Fig 5) or chin (Fig 6) touching floor. Breathe out. Complete 3-5 breathing rounds.
2. Turn head to the right with left cheek on floor. Breathe out. Complete 3-5 breathings.(Fig 7)

Fig 8

Fig 9

Fig 10

Fig 11

Fig 12

Fig 13

Fig 14

Fig 15

3. Turn head to left with right cheek on floor. Breathe out. Complete 3-5 breathing rounds.
4. Return to Fig 5.

Lying on tummy – *Parshva shavasana*

1. Lie flat on tummy. Spread arms in line with shoulders, palms down. (Fig 8)
2. Bend elbows at right angles, right arm pointing upwards, palm down and left arm pointing downwards, palm up. (Fig 9)
3. Keep the left leg straight. Bend right knee at right angle. Point toes down. Breathe out. (Fig 9)
4. Complete 3-5 breathing rounds. Repeat on opposite side.

Crocodile posture I – *Makarasana I*

1. Lie on tummy, legs apart, toes pointing outwards.
2. Cross arms and rest forehead on arms. Breathe out. (Fig 10)
3. Complete 3-5 breathing rounds.
4. Turn head to right. Breathe out.
5. Complete 3-5 breathing rounds. (Fig 11)
6. Turn head to left and repeat.

Staff posture – *Dandasana*

1. Sit upright with legs stretched forwards. (Fig 12)
2. Keep arms alongside body, palms down. Breathe out.
3. Complete 3-5 breathing rounds.

Easy Posture – *Sukhasana*

1. Sit upright with legs stretched, arms resting on thighs. (Fig 13)
2. Bend left knee and bring left leg under right thigh. (Fig 14)
3. Bend right knee and bring right leg under the left leg.
   Keep hands on knees, palms up or down. Breathe out. (Fig 15)
5. Complete 3-5 breathing rounds.

Fig 16

Fig 17

Fig 18

Fig 19

Fig 20

Fig 21

Fig 22

Fig 23

Thunderbolt posture I – *Vajrasana*

1. Kneel down on knees, toes flat. (Fig 16)
2. Slowly sit on heels. Rest palms on knees or thighs. Breathe out. (Fig 17)
3. Complete 3-5 breathings rounds.

Warrior posture – *Veerasana*

1. Kneel down, knees together, legs apart, toes flat. (Fig 18)
2. Slowly sit between heels. (Fig 19)
3. Rest palms on thighs or knees. Breathe out. (Fig 20)
4. Complete 3- 5 breathings rounds.

Deep diaphragmatic breathing – Sitting

1. Sit in Staff, Easy, Thunderbolt or Warrior posture. Place one hand on chest and the other on abdomen. Breathe out. (Fig 21)
2. Slowly breathe in deeply through the nostrils.
3. Breathe out slowly through nostrils.
   Breathe to the ratio 1:2. Breathing in 1: Breathing out 2. Only the hand on the abdomen should move while breathing. The other hand remains motionless.
4. Complete 3-5 breathing rounds.

CORPSE POSTURE FOR 5 MINUTES.

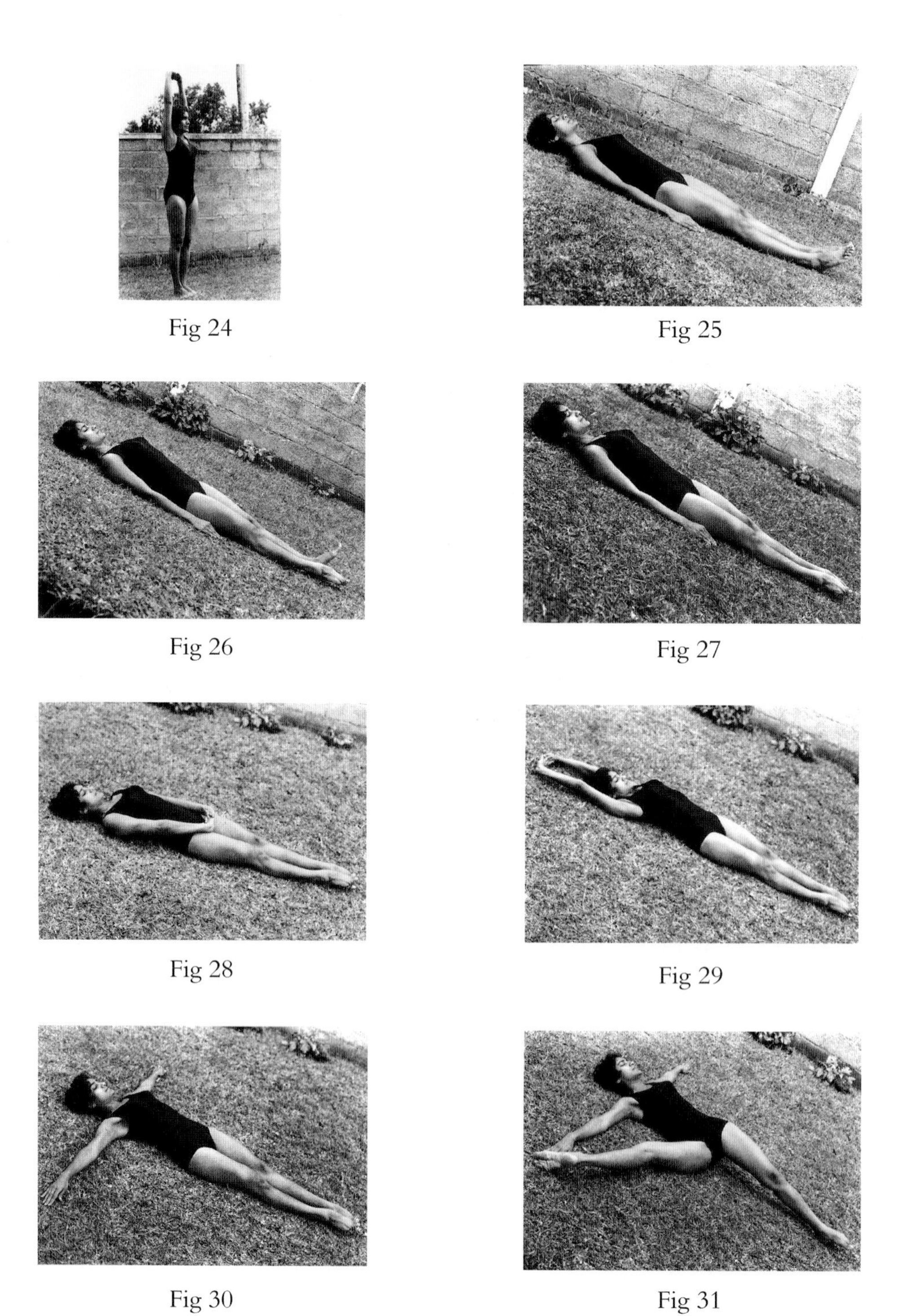

Fig 24

Fig 25

Fig 26

Fig 27

Fig 28

Fig 29

Fig 30

Fig 31

SECOND WEEK

These first two weeks are devoted entirely to relaxation and deep breathing. This week we are going to learn simple stretching postures. It will help having more awareness of ourselves, master our breathing and synchronise it with mental repetition.

We will notice how stiff and tense our body has grown over the years. How our muscles are tight and rusty. Now we are convinced we can recover some part, if not all of our youthful body. These postures with their lateral, upward and downward stretches will ease all bodily tension. They will limber up the whole body.

The movements are executed slowly with grace, never briskly. Try within limits. Do not overstrain. If found difficult in the beginning, keep on trying during later weeks until mastered. Do not be disappointed by your first attempts. Each day will bring its share of flexibility.

Yoga is known to ease stress and give peace of mind. It will also increase our levels of concentration. And this will last forever!

RELAXATION POSTURES, STANDING

Mountain Posture – *Tadasana*

1. Stand straight, legs together, arms alongside body. (Fig 22)
2. Relax all bodily tension. Distribute body weight on both legs, evenly. Breathe out.
3. Breathe in, expanding abdomen only. Breathe out, pulling abdomen in. Breathe to the ratio 1:2. Breathing in 1: Breathing out 2. To help control breathing, place hands on chest and tummy. Only the hand on the tummy should move.
4. Complete 3-5 breathing rounds.

While performing postures sitting, standing or lying on tummy, Corpse posture may be replaced by any of the above relaxation postures, always with deep diaphragmatic breathing.

STRETCHING POSTURES

Palm tree posture – *Talasana I*

1. Stand with legs together. (Fig 22). Breathe out.
2. Breathing in, raise arms above head. Stretch upper part of body from waist to maximum. Breathe out slowly. (Fig 23)
3. Maintain posture. Complete 3-5 breathing rounds. May be gradually increased to 10 breathing rounds.
4. Lower arms alongside body, breathing out. (Fig 22)
5. Relax.

Palm tree posture – *Talasana II*

1. Stand with legs together. Breathe out. (Fig 22).
2. Breathing in, cross fingers, raise arms above head, palms up. Stretch upper part of body from waist to maximum. Breathe out slowly. (Fig 24)
3. Maintain posture. Complete 3-5 breathing rounds. May be gradually increased

to 10 breathing rounds.
4. Breathing out, unlock fingers and lower arms alongside body. (Fig 22)
5. Relax.

Legs stretching – *Padapawanmuktasana*

This posture is best performed before sleep and upon waking up.

1. Lie flat on back, arms alongside body, palms down, toes pointing up. Breathe out. (Fig 25)
2. Breathing in, stretch right leg from hip to toes. (Fig 26)
3. Hold tension for 3-5 breathing rounds. Breathe out.
4. Relax.
5. Repeat with left leg.
6. Relax.
7. Repeat, stretching both legs together. (Fig 27)
8. Relax.

Stick posture – *Yeshtikasana*

1. Lie flat on back, legs together. (Fig 25)
2. Cross fingers, turning palms outwards. Breathe out. (Fig 28)
3. Breathing in, slowly raise arms to touch the floor behind the head. Breathe out. (Fig 29)
4. Breathing in, stretch the trunk and limbs to maximum, toes pointing down. Breathing out, relax tension slowly.
5. Maintain posture. Complete 3-5 breathing rounds.
6. Breathing out, lower arms and unlock fingers. (Fig 25)
7. Relax.

Fig 32

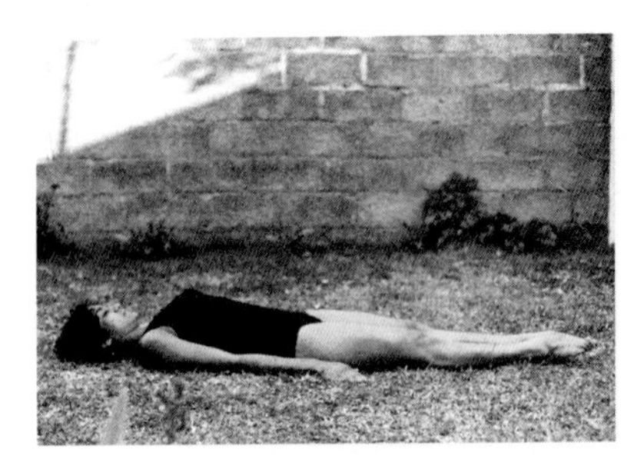

Fig 33

Fig 34

Fig 35

Fig 36

Fig 37

Fig 38

Fig 39

Sideways Stretch I – *Gokarasana I*

1. Lie down, legs together, arms spread in line with shoulders, toes pointing down. Breathe out. (Fig 30).
2. Breathing in, stretch right leg sideways to touch right hand. (Fig 31)
3. Breathing out, lower leg to starting position. (Fig 30)
4. Repeat 3 times.
5. Repeat with left leg.
6. Relax.

Sideways Stretch II – *Gokarasana II*

1. Lie down, legs together, arms spread in line with shoulders, toes pointing down. Breathe out. (Fig 30).
2. Breathing in, stretch both legs sideways as far apart as possible. (Fig 32)
3. Maintain posture. Complete 3-5 breathing rounds.
4. Breathe out brings legs together. (Fig 30)
5. Relax.

Knee bend I – *Ardha Pawanmuktasana*

1. Lie flat on back, arms alongside body, palms down, toes pointing down. Breathe out. (Fig 33)
2. Breathing in, bend right knee to rest on chest. Catch knee with hands. Breathe out. (Fig 34)
3. Maintain posture. Complete 3-5 breathing rounds.
4. Breathing out, stretch leg. (Fig 33)
5. Repeat bending left knee.
6. Relax.

Cat posture – *Bitilasana*

1. Kneel down, legs shoulder width apart, toes flat. Bend forward to rest palm on floor, in line with knees. Breathe out. (Fig 35)
2. Breathing in, arch back raising head as far as possible. (Fig 36)

3. Breathe out, pull tummy in. Round back and lower head. (Fig 37)
4. Repeat 3-5 times.
5. Sit back on heels. (Fig 38)
6. Relax.

Fish shake – *Magarasana*

1. Lie flat on back with arms alongside the body and legs stretched. (Fig 33)
2. Breathe out. Shake simultaneously body and limbs. (Fig 39).
3. Breathe in.
4. Repeat 2-3 times.

CORPSE POSTURE FOR 5 MINUTES

THIRD WEEK

Meditation is best done sitting on the floor. Sit on a small rug or mat. Choose a posture in which we can sit comfortably for at least one hour. In the beginning, 10-15 minutes breathing and meditation will be sufficient. We will be surprised how very fast we can learn sitting in Lotus posture.

It is important during meditation to synchronise deep yogic breathing with repetition of some sacred word. We shall be using 'OM' throughout the course. The results of yogic breathing and repetition of a sacred word are tremendous. They help attract an increased amount of prana. Once rhythmic action of prana is established it relaxes the body, calms the mind and helps deepen meditation.

MEDITATIVE POSTURES

Easy posture – ***Sukhasana*** (Fig 15)

Thunderbolt posture – ***Vajrasana*** (Fig 17)

Warrior posture – ***Veerasana*** (Fig 19)

Fig 40

Fig 41

Fig 42

Fig 43

Fig 44

Fig 45

Fig 46

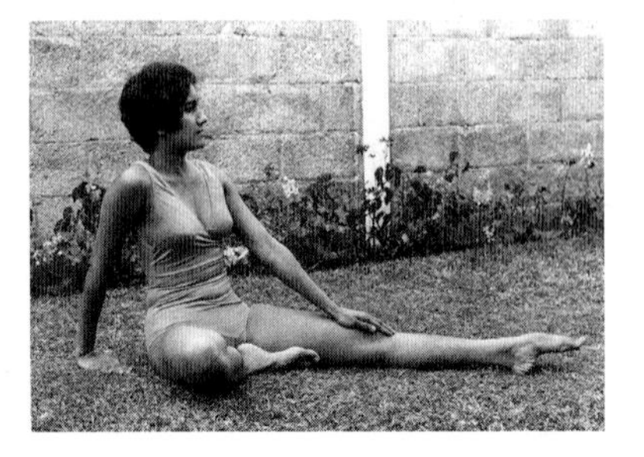

Fig 47

Lotus Posture – *Padmasana*

1. Sit upright, legs stretched. (Fig 40)
2. Bend right knee. Cross right foot over left thigh. (Fig 41)
3. Cross left foot over right thigh. (Fig 42)
4. Rest palms on knees. Breathe out. (Fig 43)
5. Hold posture for 3-5 breathing rounds.
6. Breathe out. Unlock legs and stretch them out slowly. (Fig 40)
7. Repeat bending left knee first.
8. Relax.

Half Lotus Posture – *Ardhpadmasana I*

1. Sit upright, legs stretched. (Fig 40)
2. Bend right knee. Cross right foot over left thigh. Rest palms on knees. Breathe out. (Fig 44)
3. Hold posture for 3-5 breathing rounds.
4. Breathe out. Straighten right leg. (Fig 40)
5. Repeat bending left knee.
6. Relax.

Half Lotus Posture – *Ardhpadmasana II*

1. Sit upright, legs stretched. (Fig 40)
2. Bend left knee and bring left leg under right thigh. (Fig 45)
3. Bend right knee and bring right leg over the left leg. Keep hands on knees, palms up or down. Breathe out. (Fig 46)
4. Hold posture for 3- 5 breathing rounds.
5. Breathe out. Stretch legs.(Fig 40)
6. Repeat bending right knee first.
7. Relax

Auspicious Posture – *Swastikasana*

1. Sit upright, legs stretched. (Fig 40)

Fig 48

Fig 49

Fig 50

Fig 51

Fig 52

Fig 53

Fig 54

Fig 55

2. Bend right knee, right foot close to body, toes against left thigh. (Fig 47)
3. Cross left leg over right leg. (Fig 48)
4. Gently insert left toes in between right thigh and calf, heel in contact with body. Keep hands on knees, palms up or down. (Fig 49) Breathe out.
5. Hold posture for 3-5 breathing rounds.
6. Breathe out. Stretch legs. (Fig 40)
7. Repeat bending left knee first.
8. Relax.

Perfect Posture – *Siddhasana*

1. Sit upright, legs stretched. (Fig 40)
2. Bend right knee and bring right foot close to body, toes against left thigh. (Fig 50)
3. Cross left leg over right (Fig 51), left toes tucked in between right thigh and calf. (Fig 52)
4. Keep hands on knees, palms up or down. (Fig 52) Breathe out.
5. Hold posture for 3-5 breathing rounds.
6. Breathe out. Stretch legs. (Fig 40)
7. Repeat bending left knee first.
8. Relax.

Complete Yogic Breathing

1. Sit lotus, thunderbolt, warrior or easy posture. Rest palms on knees.
2. Slowly breathe in through nostrils, first expanding abdomen (Fig 53) then chest (Fig 54) finally raising shoulders (Fig 55). All three stages are performed in a smooth continuous movement.
3. Exhale slowly, first drawing in the abdomen (Fig 56) then lowering chest (Fig 57), last lowering shoulders (Fig 58). This completes one breathing round.
4. Complete 3-5 breathing rounds. Breathe to the ratio 1:2. Breathing in 1: Breathing out 2.

Fig 56

Fig 57

Fig 58

MEDITATION TECHNIQUE

Choose any of the above postures where we will be most comfortable. In normal circumstances we become restless and fidgety after a few minutes of sitting. Our aim is to be so comfortable that we no longer feel the body. Look ahead to be sitting for a good 15 minutes, lost in deep yogic breathing and mental repetition. We always use 'OM'. It may be any other sacred word of your choice. Meditation should always be practiced with deep yogic breathing.

While sitting, rest the hands on the knees, palms up with the tips of index and thumb just touching. Keep the other fingers straight. Keep the spine very straight. Close eyes. Breathe out. Breathing in, slowly and deeply, mentally repeat 'OM' once. Pause. Breathing out, slowly and deeply, mentally repeat 'OM', 'OM'. Keep the mental repetition for the whole time you meditate. Always breathe to the ratio 1:2. Breathing in 1: Breathing out 2. Synchronise perfectly breathing and repetition of 'OM'. We will gradually reach a state of blissful well being.

After three months, we can gradually increase meditation time by five to ten minutes. After meditation, always relax in corpse posture for some 5-10 minutes.

FOURTH WEEK

As from this week always start the day's practice with these simple postures for warming up: Palm tree posture, *Talasana I* (Fig 23) Palm tree posture, *Talasana II* (Fig 24) and Stick posture, *Yestikasana* (Fig 29). This will allow the system to be more responsive.

Yoga postures proper, begin this week. Deep yogic breathing and postures combined give tremendous results. Our body will be more responsive with muscles and joints less stiff.

There is no need to perform the complete posture at first attempt. Go to the extent you can. Even if only half way, it hardly matters. Avoid all strain. As we continue working with the body it will respond better.

Holding the posture is most important, irrespective of where you have reached in the posture. It will gradually bring elasticity to the joints, muscles and ligaments. It will add to body flexibility. Nothing is too difficult. It just needs practice.

Mental repetition brings calmness to the mind and more concentration in performance and memorizing of the postures. Holding the posture and mental repetition will draw more prana to that part of the body being exercised.

Most postures will help promote health, firmness and flexibility of the spine and joints. The back stretch and forward bending further consolidate these beneficial effects. The daily chest expansion with deep yogic breathing will attract more prana.

Many among us will have the pleasant surprise of having lost quite some kilos by now!

Fig 59

Fig 60

Fig 61

Fig 62

Fig 63

Fig 64

Fig 65

Fig 66

Pump Posture – *Urdhva Prasarita Padasana*

1. Lie on back, legs together, arms alongside body. Breathe in. (Fig 59)
2. Breathing out, gradually raise both legs (Fig 60) together until vertical. (Fig 61)
3. Maintain posture. Complete 5-10 breathing rounds.
4. Breathing in, slowly lower legs together(Fig 60) to lie on back (Fig 59)
5. Relax.

Angle balance – *Samyukta santulanasana*

1. Lie on back, legs together, arms alongside body. Breathe out. (Fig 59)
2. Breathing in, slowly raise body to sitting position with knees bent, legs off floor. Hook big toes with index fingers (Fig 62) or catch ankles (Fig 64)
3. Maintain posture. Complete 1-2 breathing rounds.
4. Breathing out, slowly straighten legs, maintaining balance. (Fig 63, 65)
5. Maintain posture. Complete 3-5 breathing rounds.
6. Breathing out, slowly lower body to lie on back. (Fig 59)
7. Relax.

Twist posture – *Vakrasana I, II*

1. Sit erect, legs stretched forwards, feet together. (Fig 66)
2. Bend left knee. Cross left leg over the right to rest sole flat on floor. (Fig 67)
3. Stretch out right arm (Fig 68) to catch left toes (Fig 69) or ankles (Fig 71) **(II)**
4. Maintain posture. Complete 1-2 breathing rounds.
5. Send left arm around back to catch knee if possible. Breathe in. (Fig 69)
6. Breathing out, twist the trunk from waist up, towards left. (Fig 70) **(I)** **(**Fig 71) **(II).**
7. Maintain posture for 3-5 breathings rounds.
8. Breathing in, twist trunk to front. (Fig 69).
9. Breathe out, undo posture. (Fig 66)
10. Repeat, this time bending right knee.
11. Relax.

Fig 67

Fig 68

Fig 69

Fig 70

Fig 71

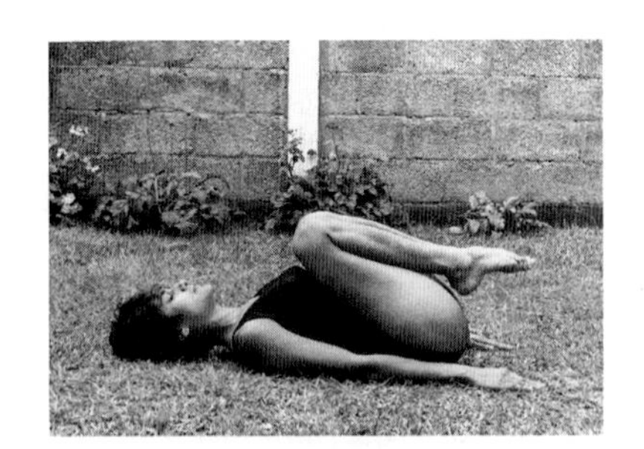

Fig 72

Fig 73

Fig 74

Knee Bend II – *Pawanmuktasana*

1. Lie on back, legs together, arms alongside body. Breathe out. (Fig 59)
2. Breathing in, bend knees to rest legs on chest. (Fig 72)
3. Maintain posture. Complete 1-2 breathing rounds.
4. Cross arms around knees. Breathe out. (Fig 73)
5. Maintain posture. Complete 3-5 breathing rounds.
6. Breathing out, undo arms, straighten legs. (Fig 59)
7. Relax.

Crescent posture – *Ardha Chandrasana*

1. Stand legs together or slightly apart, feet parallel. Breathe out. (Fig 74)
2. Breathing in, raise both arms above head. Breathe out. (Fig 75)
3. Maintain posture. Complete 1-2 breathing rounds.
4. Breathing in, bend backwards to the maximum.(Fig 76)
5. Maintain posture within limits. Complete 1-2 breathing rounds.
6. Breathing out, slowly straighten up. (Fig 75)
7. Breathing out, lower arms. (Fig 74)
8. Relax.

Crocodile Posture II – *Makarasana II*

1. Lie down, legs together, arms spread in line with shoulders, palms down. Breathe out. (Fig 77)
2. Breathe in. Bend knees, legs apart. (Fig 78)
3. Breathe out. Twist legs twice to right, left knee touching right foot, both legs to touch floor. (Fig 79)
4. Breathe in. Return to Fig 78.
5. Breathe out. Twist legs twice to left, right knee touching left foot, both legs to touch floor. (Fig 80).
6. Breathe in. Return to Fig 78.
7. Breathe out. Stretch legs. (Fig 77)
8. Relax.

Fig 75

Fig 76

Fig 77

Fig 78

Fig 79

Fig 80

Fig 81

Fig 82

Neck Roll – *Grivavartenasana*

1. Sit thunderbolt posture, palms on knees. (Fig 81)
2. Breathe out. Bend head forward. (Fig 82)
3. Breathe in. Roll head to right (Fig 83), back (Fig 84), to left (Fig 85) and forward again (Fig 82). Breathe out.
4. Breathe in and raise head. (Fig 81)
5. Repeat 3-5 times.
6. Repeat rolling head to left. Relax.

MEDITATION

CORPSE POSTURE 5-10 minutes.

Fig 83

Fig 84

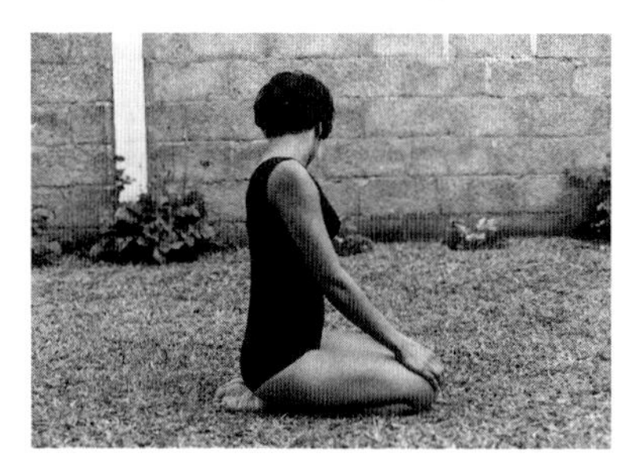

Fig 85

Fig 86

Fig 87

Fig 88

Fig 89

Fig 90

FIFTH WEEK

The new postures have to be practiced daily as usual. If there is extra time we may include, on and off, a few postures of previous weeks. Else, we may gradually increase the number of breathing rounds by one or two.

The daily chest expansion postures, the back stretches, will help reduce stiffness and increase ease of movements in all directions. Cobra postures are probably the best lung exercises. The muscles of the thorax are strengthened.

Everyone will enjoy the child posture, considered among the easiest. Nevertheless it gives tremendous relaxation to the whole system.

Catching Big Toe – *Sapta Padangusth Asana I*

1. Lie flat on back, legs together, arms alongside body. Breathe out. (Fig 86)
2. Breathing in, raise left leg. Raise left arm to hook big toe with index finger (Fig 87) or catch ankle. (Fig 88) Breathe out.
3. Maintain posture. Complete 3-5 breathing rounds.
4. Breathing in, slowly lower left arm and leg. (Fig 86)
5. Repeat with right arm and leg.
6. Relax.

Squat posture – *Malasana*

1. Stand feet together, hands folded. (Fig 89)
2. Breathe in, slowly raise on toes. (Fig 90)
3. Breathe out, slowly squat down.(Fig 91)
4. Sit on heels. Keep hands folded (Fig 92) or rest palms on knees (Fig 93)
5. Maintain posture. Complete 1-2 breathing rounds.
6. Flatten feet, place palms flat on either side of feet. Breathe out. (Fig 94)
7. Maintain posture. Complete 3-5 breathing rounds.
8. Relax.

Fig 91

Fig 92

Fig 93

Fig 94

Fig 95

Fig 96

Fig 97

Fig 98

Equestrian Posture I – ***Ashwa sanchalana asana I***

Toes raised

1. Squat on heels, place both palms on floor, arms straight. (Fig 95)
2. Send right leg backwards. Breathe out. (Fig 96)
3. Maintain posture. Complete 1-2 breathing rounds.
4. Breathing in, bend trunk backwards as far as possible. Breathe out. (Fig 97)
5. Maintain posture. Complete 3-5 breathing rounds.
6. Breathing out, bend forward to (Fig 96)
7. Bring right leg forward to rest alongside left. (Fig 95)
8. Repeat, sending left leg backwards.
9. Relax.

Equestrian Posture II- ***Ashwa sanchalana asana II***

Toes flat

1. Squat on heels, place both palms on floor, arms straight.(Fig 95)
2. Send right leg backwards, flattening right toes. Breathe out. (Fig 98)
3. Maintain posture. Complete 1-2 breathing rounds.
4. Breathing in, bend trunk backwards as far as possible. Breathe out. (Fig 99)
5. Maintain posture. Complete 3-5 breathing rounds.
6. Breathing out, bend forward to (Fig 98)
7. Bring right leg forward to rest alongside left. (Fig 95)
8. Repeat, sending left leg backward.
9. Relax.

Crescent lunge posture I – ***Anjanaya asana I***

Arms and toes raised

1. Squat on heels, place both palms on floor, arms straight. (Fig 95)
2. Send right leg backwards, keeping toes raised. Breathe out. (Fig 96)
3. Maintain posture. Complete 1-2 breathing rounds.

Fig 99

Fig 100

Fig 101

Fig 102

Fig 103

Fig 104

Fig 105

Fig 106

4. Breathing in, raise arms above head and bend trunk backwards as far as possible. Breathe out. (Fig 100)
5. Maintain posture. Complete 3-5 breathing rounds.
6. Breathing out, lower arms, place palms on floor (Fig 96)
7. Bring right leg forward to rest alongside left (Fig 95)
8. Repeat, sending left leg backwards.
9. Relax.

Crescent lunge posture II – ***Anjanaya asana II***

Arms raised and toes flat

1. Squat on heels, place both palms on floor, arms straight.(Fig 95)
2. Send right leg backwards, flattening right toes. (Fig 98)
3. Maintain posture. Complete 1-2 breathing rounds.
4. Breathing in, raise arms above head and bend trunk backwards as far as possible. Breathe out. (Fig 101)
5. Maintain posture. Complete 3-5 breathing rounds.
6. Breathing out, lower arms, place palms on floor (Fig 98)
7. Bring right leg forward to rest alongside left (Fig 95)
8. Repeat, sending left leg backwards.
9. Relax.

Palm tree posture III – *Talasana III*

1. Stand with legs shoulder width apart. (Fig 102). Breathe out.
2. Maintain posture. Complete 1-2 breathing rounds.
3. Breathing in, raise arms above head. Stretch upper part of body from waist to maximum. Breathe out slowly. (Fig 103)
4. Maintain posture. Complete 3-5 breathing rounds. May be increased gradually up to 10 breathing rounds.
5. Lower arms alongside body, breathing out. (Fig 102)
6. Relax.

Palm tree posture IV – *Talasana IV*

1. Stand with legs shoulder width apart. Breathe out. (Fig 102)
2. Maintain posture. Complete 1-2 breathing rounds.
3. Breathing in, cross fingers (Fig 104).
4. Raise arms above head, palms up. Stretch upper part of body from waist to maximum. Breathe out slowly. (Fig 105)
5. Maintain posture. Complete 3-5 breathing rounds. May be increased gradually up to 10 breathing rounds.
6. Breathing out, unlock fingers and lower arms alongside body. (Fig 102)
7. Relax.

Child posture – *Balasana*

1. Sit on heels, palms resting on knees or thighs. Breathe out (Fig 106)
2. Bend forward. Send arms backwards to rest alongside body, palms up. Rest forehead on floor. (Fig 107)
3. Maintain posture. Complete 3-5 breathing rounds.
4. Breathing in, sit up. (Fig 106)
5. Relax.

MEDITATION

CORPSE POSTURE 5-10 minutes

SIXTH WEEK

Postures already learnt in previous weeks serve as basis to more advanced ones. We will notice that by simply adding some extra movements to them, we are obtaining more and more advanced postures.

The veins are among the weakest components of the circulatory system. The leg veins are especially at risk having to support the weight of the blood in the whole body. More and more people have to stand for longer hours because of the nature of their work. They are more prone to varicose veins, a disturbing and sometimes very painful condition. The practice of reverse postures is especially beneficial in these cases.

Lion postures help those with throat ailments. Finger tension will relieve all stress in hands and fingers, excellent for our computer marathoners. We have scores of cases who got cured of longstanding finger stiffness or arthritis of the hands.

As from this week we have two more postures Palm Tree III – *Talasana III* (Fig 103) and Palm Tree IV – *Talasana IV* (Fig 105) that can be used as warm up. We can alternate them with the previous Palm Tree postures.

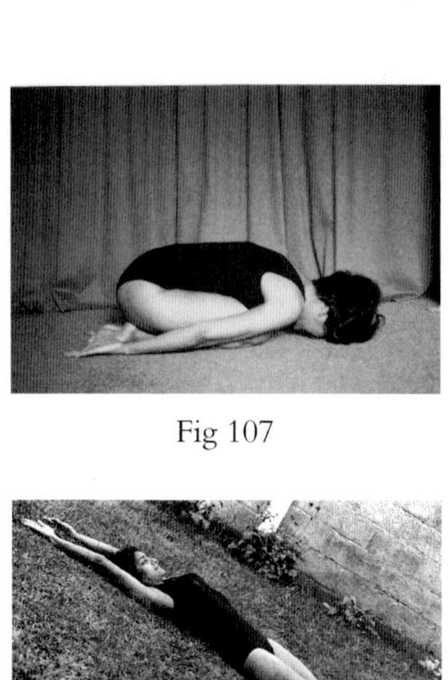

Fig 107

Fig 108

Fig 109

Fig 110

Fig 111

Fig 112

Fig 113

Fig 114

Catching both toes – ***Sapta Padangusth Asana II***

1. Lie on back, legs together, arms alongside body. Breathe out. (Fig 108)
2. Breathing in raise arms to touch floor behind head.(Fig 109)
3. Breathing out, raise both legs together. (Fig 110)
4. Maintain posture. Complete 1-2 breathing rounds.
5. Hook big toes with index (Fig 111) or catch ankles with hands. (Fig 112) Breathe out.
6. Maintain posture. Complete 3-5 breathing rounds.
7. Breathing in, slowly lower arms and legs. (Fig 109)
8. Breathing out, lower arms to (Fig 108)
9. Relax.

Reverse posture – ***Viparitakarani asana***

1. Lie on back, legs together, arms alongside body. Breathe in. (Fig 108)
2. Breathing out, raise hips, keeping knees bent. Use hands to support hips. Breathe out. (Fig 113)
3. Maintain posture. Complete 1-2 breathing rounds.
4. Breathing in, straighten forelegs towards head. Breathe out. (Fig 114)
5. Maintain posture. Complete 3-5 breathing rounds.
6. Breathing out, lower arms and lower legs towards head. (Fig 115)
7. Breathing in, slowly lower back to floor (Fig 116) then legs. (Fig 108)
8. Relax.

Fig 115

Fig 116

Fig 117

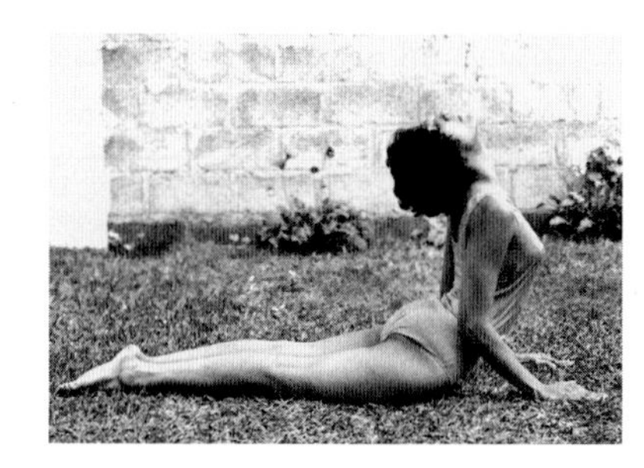

Fig 118

Fig 119

Fig 120

Fig 121

Fig 122

Cobra posture – ***Bhujanga Asana***

1. Lie flat on tummy, forehead on floor. Place palms on floor, level with armpits. Breathe out. (Fig 117)
2. Breathing in, slowly raise head. Keep elbows bent, lift shoulders and trunk as far as possible. Breathe out. (Fig 118)
3. Maintain posture. Complete 3-5 breathing rounds.
4. Breathe out, lower body to rest forehead on floor. (Fig 117)
5. Relax.

Swan posture – ***Hansasana***

1. Lie flat on tummy, forehead on floor. Place palms on floor, level with armpits. Breathe out. (Fig 117)
2. Breathe in. Keeping palms and knees on floor press on hands (Fig 119) and raise body backwards (Fig 120) to sit on heels (Fig 121)
3. Place chin (Fig 121) or forehead (Fig 122) between arms on floor. Breathe out.
4. Maintain posture. Complete 3-5 breathing rounds.
5. Breathing in raise body and sit back on heels, hands on knees. (Fig 123)
6. Relax.

Thunderbolt posture II– ***Vajrasana II***

1. Sit on heels, palms resting on knees. Breathe out. (Fig 123)
2. Cross fingers (Fig 124). Breathing in, raise arms above head.(Fig 125)
3. Maintain posture. Complete 3-5 breathing rounds.
4. Breathing out, lower arms to rest on knees. (Fig 123)
5. Relax.

Fig 123

Fig 124

Fig 125

Fig 126

Fig 127

Fig 128

Fig 129

Fig 130

Fish posture I – ***Matsyasana I***

1. Sit thunderbolt. Breathe out. (Fig 123)
2. Lean backwards, place palms on floor, fingers pointing forward. Breathe out. (Fig 126)
3. Breathe in, lower body. Place one elbow after another on floor. (Fig 127)
4. Slowly slide forearms towards body. Rest crown of head on floor, arms on thighs. Breathe out. (Fig 128)
5. Maintain posture. Complete 3-5 breathing rounds.
6. Breathe out. Place palms on either side of head. Lower head. Breathe in. (Fig 129)
7. Breathe out. Place elbows on floor. (Fig 127) With help of arms raise body to sit up. (Fig 126).
8. Relax, sitting thunderbolt. (Fig 123)

Lion posture I – ***Simhasana I***

1. Sit thunderbolt posture, palms on knees. Breathe out. (Fig 123)
2. Spread fingers freely, hands just touching knees. (Fig 130)
3. Breathe in, hold. Pull out tongue and press chin against throat. (Fig 130) Fix gaze between eyebrows. (Fig 131).
4. Hold posture as long as comfortable, 5-30 seconds.
5. Pull in tongue. Raise chin. Relax fingers. Breathe out. (Fig 123)
6. Relax.

Lion Posture II – ***Simhasana II***

1. Sit thunderbolt posture, palms on knees. Breathe out. (Fig 123)
2. Spread fingers and press palms hard against knees. (Fig 132)
3. Breathe in, hold. Pull out tongue and press chin against throat. Fix gaze between eyebrows. (Fig 133).
4. Hold posture as long as comfortable, 5-30 seconds.
5. Pull in tongue. Raise chin. Relax fingers. Breathe out. (Fig 123)
6. Relax.

Fig 131

Fig 132

Fig 133

Fig 134

Fig 135

Fig 136

Fig 137

Fig 138

Finger tension – *Hastapawanmukta*

1. Sit thunderbolt posture. Breathe out. (Fig 123)
2. Breathing in, tense fingers, close fists slowly, maintaining tension in both forearm and fingers. (Fig 134)
3. Breathing out, open fists, all the while maintaining tension in forearm and fingers. (Fig 135)
4. Repeat 5-10 times.
5. Relax tension in forearms and fingers.

MEDITATION

CORPSE POSTURE 5-10 minutes

Fig 139

Fig 140

Fig 141

Fig 142

Fig 143

Fig 144

Fig 145

Fig 146

SEVENTH WEEK

Shoulder stand will reinforce the beneficial effects obtained from last week's reverse postures. It is an advanced version and its effects are more powerful. Follow carefully all the steps and instructions for every posture. You will be amazed how well you can perform them when done the proper way.

There is no rigid sequence in the performance of the weekly series. It is better to change from one type of posture to another, sitting, standing or lying down. This gives the body some extra movements.

Except at the end of the daily series, there is no need to relax in corpse posture after every posture. After a sitting posture choose a sitting relaxation posture. Standing postures can be followed by standing relaxation one. Lying on back or tummy can be followed by corresponding lying relaxation postures.

Face exercises will bring an additional glow, with all the novel gentle movements and massage. A plus for all, young and old alike!

Half Cobra I – *Ardha Bhujanga asana I*

1. Kneel down. (Fig 136)
2. Bring right leg forward. Breathe in. (Fig 137)
3. Maintain posture. Complete 1-2 breathing rounds.
4. Breathing out, keep spine erect, lower body until palms are flat on floor (Fig 138) or fingers touch floor (Fig 139).
5. Maintain posture. Complete 3-5 breathing rounds.
6. Raise body to Fig 137.
7. Kneel down. (Fig 136)
8. Repeat with left leg forward.
9. Relax.

Fig 147

Fig 148

Fig 149

Fig 150

Fig 151

Fig 152

Fig 153

Fig 154

Twisting Cross – *Gokarnasana*

1. Lie on back, arms spread in line with shoulders, palms down. Breathe out. (Fig 140)
2. Breathe in. Turn face to left. (Fig 141)
3. Breathing out, cross left leg over right leg and drag it on floor to reach right arm. Try to catch right big toe with index. Keep legs perfectly straight and shoulders flat on floor.(Fig 142)
4. Maintain posture. Complete 3-5 breathing rounds.
5. Breathing in, turn head straight, lower leg. (Fig 140)
6. Turn face to the right and repeat with right leg crossed over left.
7. Relax.

Forward Bend – *Paschimotanasana I*

1. Lie on back, legs together, arms alongside body. Breathe out. (Fig 143)
2. Breathing in, raise arms to touch floor behind head. (Fig 144)
3. Complete 1-2 breathing rounds, stretching both arms and legs.
4. Breathing out, in one movement sit up (Fig 145) and bend forward touching knees with forehead. (Fig 146)
5. Keeping knees straight, elbows resting on floor, catch big toes with fingers (Fig 146) or grasp ankles with hands.(Fig 147)
6. Maintain posture. Complete 3-5 breathing rounds.
7. Breathing in, in one movement, raise arms (Fig 145) bend backward to lie on floor. (Fig 144)
8. Breathing out lower arms to the sides. (Fig 143)
9. Relax.

Forward Bend – *Paschimotanasana II*

Sitting

1. Sit erect, legs stretched forwards, feet together, palms on knees. Breathe out. (Fig 148)

Fig 155

Fig 156

Fig 157

Fig 158

Fig 159

Fig 160

Fig 161

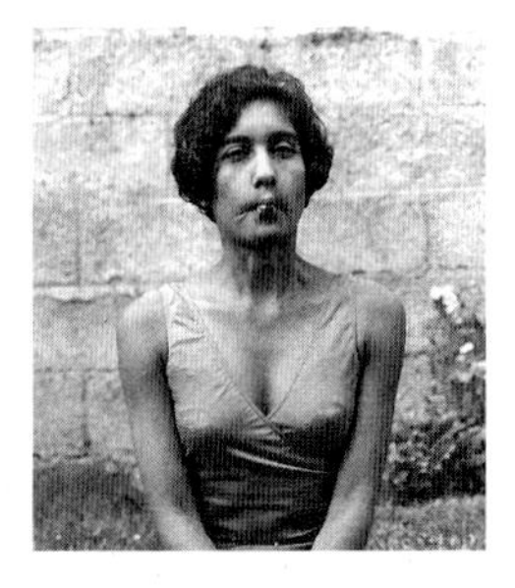

Fig 162

2. Breathing in, raise both arms above head. (Fig 149) Complete 1-2 breathing rounds.
3. Breathing out, bend forward to touch knees with forehead, if possible, and catch big toes. (Fig 146)
4. Otherwise, breathing out, bend slowly forward to catch big toes with fingers or ankles with hands (Fig 150)
5. Maintain posture. Complete 3-5 breathing rounds.
6. Breathing in, raise arms. Complete 1-2 breathing rounds. (Fig 149)
7. Breathing out lower arms to rest palms on knees. (Fig 148)
8. Relax.

Wedge Posture – *Purvottanasana*

1. Sit erect, legs stretched forwards, feet together, palms on knees. Breathe out. (Fig 148)
2. Place palms on floor behind hips, fingers pointing forwards. (Fig 151)
3. Breathe in. Bend knees to rest soles of feet flat on floor. (Fig 152)
4. Maintain posture. Complete 1-2 breathing rounds.
5. Breathing out, raise body, stretching arms and legs. Bend head back. (Fig 153)
6. Maintain posture. Complete 3-5 breathing rounds.
7. Breathing out, slowly lower body to sit on floor. (Fig 151)
8. Relax.

Shoulder stand Posture – *Sarvangasana*

1. Lie on back, legs together, arms alongside body, palms down. (Fig 143)
2. Breathing out, raise legs towards head. Place hands against back to support hips.(Fig 154)
3. Maintain posture. Complete 1-2 breathing rounds.
4. Breathing in, raise legs perpendicular to floor, toes pointing up. Keep chin pressed against chest. Breathe out. (Fig 155)
5. Maintain posture. Complete 3-5 breathing rounds.
6. Breathing out, lower arms to floor, lower legs towards head. (Fig 156)
7. Breathing in, slowly lower back first (Fig 157) then legs to floor. (Fig 143)
8. Relax.

Fig 163

Fig 164

Fig 165

Fig 166

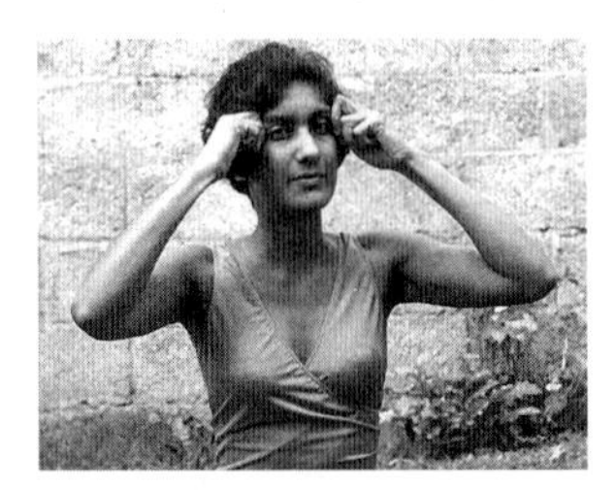

Fig 167

Fig 168

Fig 169

Fig 170

Fig 171

Plough Posture – *Halasana*

1. Lie on back, legs together, arms alongside body, palms down. (Fig 143)
2. Breathing out, raise legs to touch floor behind head with toes. (Fig 158). If unable to touch floor, hold legs above head (Fig 156).
3. Maintain posture. Complete 3-5 breathing rounds.
4. Breathing in, slowly lower back first, (Fig 157) then legs to floor (Fig 143).
5. Relax.

Face Exercises

1. Sit anywise for face exercises.
2. Puff both cheeks hard. Hold for 2-3 breathing rounds. (Fig 159)
3. Puff both cheeks hard. Apply gentle pressure on each cheek, in turn, with fingers, pushing air the opposite side. Breathe normally. (Fig 160, 161)
4. Draw corners of mouth inside until they touch each other. Hold for 2-3 breathing rounds. (Fig 162)
5. Move chin muscles in circular motion with help of fingers, first clockwise, then anticlockwise. Breathe normally. (Fig 163)
6. Insert index fingers at both corners of mouth. Breathe out. (Fig 164) Breathing in, pull gently sideways. (Fig 165) Breathing out, relax. Repeat 2-3 times.
7. Gently take hold of skin at temple between back of index and middle fingers. (Fig 166) Firmly move up (Fig 167) and down (Fig 168) without raising fingers. Breathe normally. Repeat 2-3 times.
8. Massage nape of neck (Fig 169), behind ears (Fig 170), and at temples (Fig 171), moving fingers in circular motion, clockwise and anticlockwise. Breathe normally.

MEDITATION

CORPSE POSTURE, 5-10 minutes

Fig 172

Fig 173

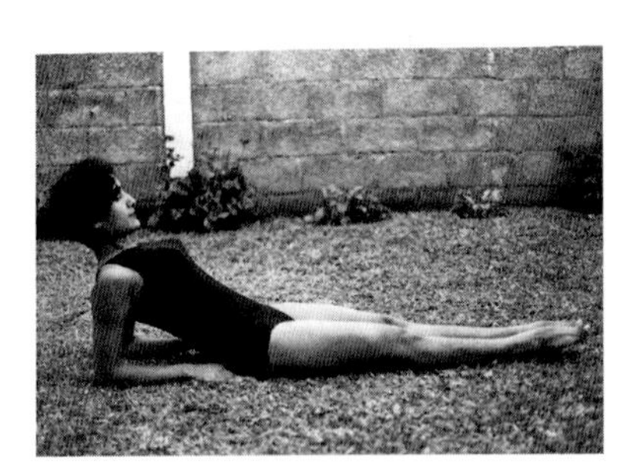

Fig 174

Fig 175

Fig 176

Fig 177

Fig 178

Fig 179

EIGHTH WEEK

By now most of us have experienced the beneficial changes in our body, habits and mental outlook. We should be noticing marked improvements in our daily life; better appetite, sound sleep, reduced stress, more discipline in sleeping habits, etc.

Locust postures are the quintessence of perfection as lung exercises. Hence the importance of maintaining the posture, together with deep breathing. The respiratory muscles are thoroughly exercised and strengthened by this powerful combination.

We are now able to cope better with the postures. Anytime is convenient for practice, so long as we can devote some 30-45 minutes undisturbed. Perform in peace for maximum results. Anytime during the day we can fit in the face exercises. They will work all facial muscles and reduce or prevent wrinkle formation. These simple exercises go a long way to ease tension.

Fish Posture II – *Matsyasana II*

1. Sit upright with legs stretched. (Fig 172)
2. Lean backward. Cross forearms behind back (Fig 173) or leave arms alongside body. (Fig 174) Breathe out.
3. Breathe in, arch back. Slide forearms towards body until head rests on floor. Breathe out. (Fig 175, 176)
4. Maintain posture. Complete 3-5 breathing rounds.
5. Breathe in. Pressing with palms on either side of head, fingers towards body, lower head to floor. (Fig 177)
6. Breathe out. Lower arms to the side.
7. Relax.

Half Cobra II- *Ardha Bhujanga Asana II*

1. Kneel down. (Fig 178)
2. Place right leg forward. Rest right palm on knee. Breathe in. (Fig 179)
3. Breathe out slowly, lower trunk. Twist to left, left arm to touch floor alongside left knee. (Fig 180, 181)

Fig 180

Fig 181

Fig 182

Fig 183

Fig 184

Fig 185

Fig 186

Fig 187

4. Maintain posture. Complete 3 breathing rounds.
5. Breathing in, raise body to Fig 179.
6. Kneel down. (Fig 178)
7. Repeat with left leg forward. Relax.

Cow Posture – *Gomukhasana*

1. Sit upright with legs stretched. (Fig 172)
2. Bend left leg to sit on left heel.(Fig 182)
3. Cross right leg over left thigh, right heel on floor. Breathe out.
4. Breathe in. Bend left arm over left shoulder. Bend right arm behind back to catch left arm. Interlock fingers (Fig 183) or place palms flat against each other (Fig 184) Breathe out.
5. Maintain posture. Complete 3-5 breathing rounds.
6. Breathe in. Repeat, bending right arm over right shoulder and left arm behind back. (Fig 185)
7. Repeat bending right leg to sit on right heel. (Fig 186)
8. Relax.

Half Locust Posture – *Ardh Salabhasana*

1. Lie on tummy, arms alongside body, palms up, chin on floor. (Fig 187)
2. Breathing in, raise right leg straight. Breathe out. (Fig 188)
3. Maintain posture. Complete 3-5 breathing rounds.
4. Breathing out, slowly lower leg. (Fig 187)
5. Repeat raising left leg.
6. Relax.

Locust Posture I – *Salabhasana I*

1. Lie on tummy, arms alongside body, palms up, chin on floor. (Fig 187)
2. Breathing in, press on hands, raise legs straight. Breathe out. (Fig 189)
3. Maintain posture. Complete 3-5 breathing rounds.
4. Breathing out, slowly lower legs. (Fig 187)
5. Relax.

Fig 188

Fig 189

Fig 190

Fig 191

Fig 192

Fig 193

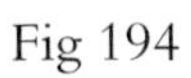

Fig 194

Fig 195

Locust Posture II– *Salabhasana II*

1. Lie on tummy, arms alongside body, palms up, chin on floor. (Fig 187)
2. Close fists, thumbs down. Breathing in, press on fists, raise legs straight. Breathe out. (Fig 190)
3. Maintain posture. Complete 3-5 breathing rounds.
4. Breathing out, slowly lower legs. (Fig 187)
5. Relax.

Locust Posture III – *Salabhasana III*

1. Lie on tummy, arms alongside body, palms under thighs, chin on floor. (Fig 191)
2. Breathing in, raise legs straight, pushing thigh up with palms. Breathe out. (Fig 192)
3. Maintain posture. Complete 3-5 breathing rounds.
4. Breathing out, lower legs.
5. Relax.

Tortoise posture – *Uttana Kurmasana*

1. Sit between heels. Palms on knees. (Fig 193)
2. Breathing in, raise arms. (Fig 194)
3. Maintain posture. Complete 1-2 breathing rounds.
4. Breathing out, bend down, arms backwards (Fig 195).
5. Rest palms against soles of feet, forehead on floor, chin between knees. (Fig 196)
6. Maintain posture. Complete 3-5 breathing rounds.
7. Breathing in, sit up. (Fig 193)
8. Relax.

MEDITATION

CORPSE POSTURE 5-10 minutes.

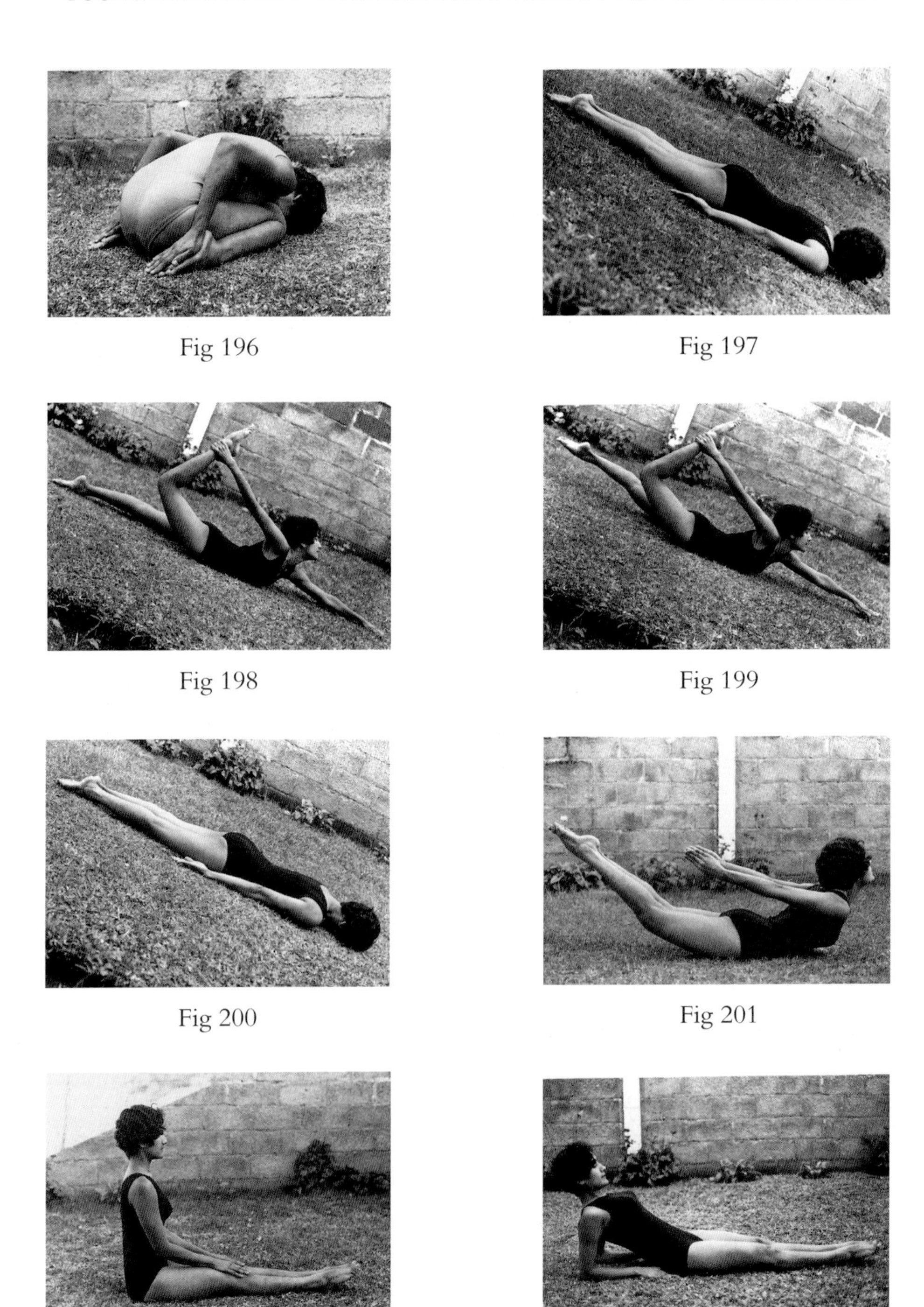

Fig 196

Fig 197

Fig 198

Fig 199

Fig 200

Fig 201

Fig 202

Fig 203

NINTH WEEK

During this week there is considerable stretching, backwards and forwards. Every part of the body is being exercised. Whether we can partly or fully manage them, we should attempt all postures from the first day. Gradually we will tackle them better and better.

Most postures will help promote firmness and flexibility of the spine and joints. This newfound suppleness will blend with strength and coordination. There will be enhanced cardiovascular fitness. We feel a growing sense of physical confidence and overall well-being.

In this computer age much of our time is spent in company of our magic screens. The eye exercises come as a soothing workout. They tone all the external muscles of the overtired eyeball and give them perfect relaxation.

Half Bow posture I- *Ardha dhanurasana I*

1. Lie on tummy, arms alongside body, palms up, chin on floor. (Fig 197)
2. Stretch left arm forward. Bend right knee and catch right ankle or toes with right hand. Breathing in, slowly raise right leg and head off floor. Keep back well arched. (Fig 198)
3. Maintain posture. Complete 3-5 breathing rounds.
4. Breathing out, lower right leg and head.
5. Repeat with left leg, right arm stretched forward.
6. Relax.

Half Bow posture II- *Ardha dhanurasana II*

1. Lie on tummy, arms alongside body, palms up, chin on floor. (Fig 197)
2. Stretch left arm forward. Bend right knee and catch right ankle or toes with right hand. Breathing in, slowly raise right leg and head off floor. Keep back well arched. Raise left arm and leg a few inches off floor. (Fig 199)
3. Maintain posture. Complete 3-5 breathing rounds.
4. Breathing out, lower arms, legs and head.
5. Repeat with left arm catching left leg, right arm stretched forward.
6. Relax.

Fig 204

Fig 205

Fig 206

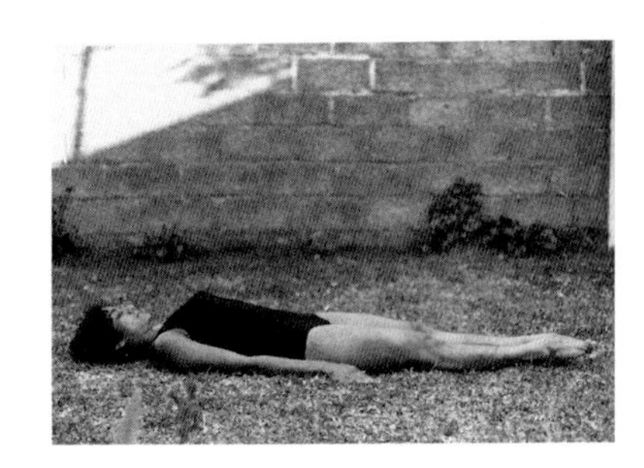

Fig 207

Fig 208

Fig 209

Fig 210

Fig 211

Locust posture- ***Purna Salabhasana***

1. Lie on tummy, arms alongside body, legs together, forehead on floor. (Fig 200)
2. Breathing in, raise head, arms and legs off floor. Keep arms parallel, shoulder width apart, legs straight and close together. Breathe out. (Fig 201)
3. Maintain posture. Complete 3-5 breathing rounds.
4. Breathing out, lower head and limbs.
5. Relax.

Fish Posture III – ***Matsyasana III***

1. Sit upright with legs stretched. Breathe out. (Fig 202)
2. Lean backward to allow forearms to rest on floor behind hips. (Fig 203)
3. Maintain posture. Complete 1-2 breathing rounds.
4. Breathe in, arch back. Slide forearms towards body until head touches floor. Breathe out. (Fig 204)
5. Breathe in. Place arms a little away from body, palms up. Bend knees, join soles of feet together. Breathe out. (Fig 205)
6. Maintain posture. Complete 3-5 breathing rounds.
7. Breathe out. Straighten legs.
8. Breathe in. Pressing with palms on either side of head, fingers towards body, lower head to floor. (Fig 206)
9. Breathe out. Lower arms alongside body. (Fig 207)
10. Relax.

Head to knee posture – ***Janushirasana***

1. Sit upright, legs stretched, feet together, hands on knees. Breathe out. (Fig 202)
2. Bend right leg, to rest foot against left thigh. Breathe in. (Fig 208)
3. Maintain posture. Complete 1-2 breathing rounds.
4. Breathing out, bend forward and hook left big toe with index fingers, (Fig 209) or catch ankles (Fig 210). Touch knees with forehead.
5. Maintain posture. Complete 3-5 breathing rounds.

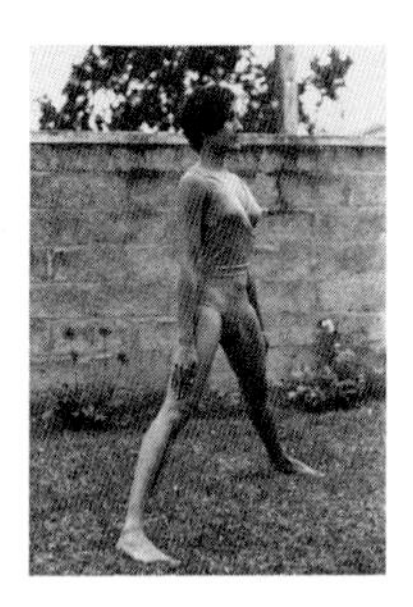

Fig 212

Fig 213

Fig 214

Fig 215

Fig 216

Fig 217

Fig 218

Fig 219

6. Breathe in, raise body and straighten legs. (Fig 202)
7. Repeat, bending left leg.
8. Relax.

Limb spread posture – *Utthita hastapadasana*

1. Stand legs together, arms alongside body. (Fig 211)
2. Breathing in spread legs some one metre apart. (Fig 212)
3. Maintain posture. Complete 1-2 breathing rounds.
4. Breathing in. Spread arms in line with shoulders.(Fig 213)
5. Maintain posture. Complete 3-5 breathing rounds.
6. Breathing out, lower arms. Bring legs together. (Fig 211)
7. Relax.

Limbs spread posture, Lateral – *Parshva hastapadasana*

1. Stand legs together, arms alongside body. (Fig 211)
2. Breathing in, spread legs some one metre apart. (Fig 212)
3. Maintain posture. Complete 1-2 breathing rounds.
4. Breathing in. Spread arms in line with shoulders. (Fig 213)
5. Turn left leg sideways, at right angle.(Fig 214)
6. Maintain posture. Complete 3-5 breathing rounds.
7. Breathing out, lower arms. Bring legs together. (Fig 211)
8. Relax.

Tree Posture – ***Dhruvasana***

1. Stand legs together, arms alongside body. Breathe out. (Fig 211)
2. Breathe in. Bend right knee, half lotus. Fold arms in front of chest. Breathe out. (Fig 215)
3. Maintain posture. Complete 3-5 breathing rounds.
4. Breathing out, lower right leg and lower arms. (Fig 211)
5. Repeat, bending left knee.
6. Relax.

Fig 220

Fig 221

Fig 222

Fig 223

Fig 224

Fig 225

Fig 226

Fig 227

Stork Posture I, II, III, IV– *Padhastasana I, II, III, IV*

1. Stand, legs together, arms alongside body. (Fig 211)
2. Breathing in, raise arms above head, elbows straight. (Fig 216)
3. Maintain posture. Complete 1-2 breathing rounds.
4. Breathing out, keeping head between arms, bend forward (Fig 217) to place palms flat on floor. (Fig 218) **(I),** catch big toes with index fingers (Fig 219) **(II)**, or catch ankles (Fig 220) **(III)**.
5. If unable to catch either toes or ankles, try to touch toes with fingers. (Fig 221) **(IV)**
6. Maintain posture. Complete 3-5 breathing rounds.
7. Breathing in, straighten up, keeping head between arms.
8. Breathing out, lower arms to sides. (Fig 211)
9. Relax.

Eye exercises – *Netra Vyayamam*

1. Sit Lotus, Easy, Thunderbolt or anywise, facing a wall some 10 ft away. (Fig 222)
2. Keep the head straight. Only the eyes should move slowly. Fix a point on the wall, level with eyes. Breathe in. (Fig 222)
3. Breathing out, slowly lower eyes down the wall and floor towards legs. (Fig 223)
4. Breathing in return eyes to starting point. (Fig 222)
5. Repeat 3-5 times.
6. Breathing out move eyes up the wall and ceiling towards head. (Fig 224)
7. Breathing in return eyes to starting point. (Fig 222)
8. Repeat 3-5 times.
9. Breathe out. Move eyes sideways to right (Fig 225).
10. Breathing in return eyes to starting point. (Fig 222)
11. Repeat 3-5 times.
12. Breathe out. Move eyes sideways to left (Fig 226).
13. Breathing in return eyes to starting point. (Fig 222)
14. Repeat 3-5 times.
15. Rest eyes by winking 5-10 times with closed eyelids, breathing normally.

16. Fix starting point on the wall, level with eyes. (Fig 222). Breathe in.
17. Breathing out, lower eyes down the wall and floor towards legs. (Fig 223)
18. Breathing in, move eyes half a circle, to the right (Fig 227) then above head. (Fig 228)
19. Breathing out, move eyes to the left (Fig 229) and down, (Fig 230) to complete the circle.
20. Repeat 3-5 times anticlockwise and 3-5 times clockwise.
21. Rest eyes by winking 5-10 times with closed eyelids, breathing normally.

MEDITATION

CORPSE POSTURE, 5-10 minutes

Fig 228

Fig 229

Fig 230

TENTH WEEK

The hazards of modern life style have robbed man of his chances to perfect health, peace and happiness. It is our sacred duty to do our utmost to achieve them. Yoga and meditation offer us the opportunity of a lifetime. It is up to us to make the best of this exceptional privilege.

Every posture is truly galvanizing to the whole body, with considerable stimulation of all systems. Muscle tension will be greatly reduced. It will ease off much bodily and mental tension. Concentrate on every single posture, whether easy to perform or comparatively more difficult to tackle. Practice makes perfect. It will come in its own time. Patience is gold.

Scalp exercises are part of this week's series. They are splendid for stress, headache and hair problems. Performed at bedtime they are excellent to induce sound sleep.

We should find time to daily practice the eye exercises. The two new wonderful eye exercises are a blessing in disguise for they tone both external and internal eye muscles. They are indeed unique and we should not neglect them. We can do all together or separately any time during the day or before retiring to bed.

Fig 231

Fig 232

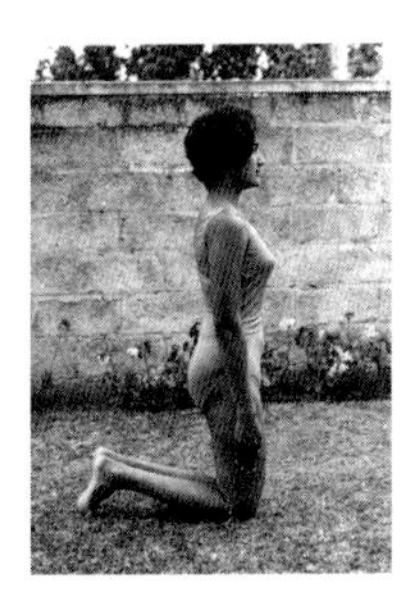
Fig 233

Fig 234

Fig 235

Fig 236

Fig 237

Fig 238

Palm tree posture V – *Talasana V*

1. Stand legs together, fingers interlocked in front. (Fig 231)
2. Breathing in, raise on toes, with arms above head. Stretch body and limbs to maximum. Breathe out, release tension. (Fig 232)
3. Keep on toes. Complete 3-5 breathing rounds.
4. Breathing out, lower arms and flatten feet.
5. Relax.

Camel Posture – *Ustrasana*

1. Kneel down, toes raised, arms alongside body. Breathe out. (Fig 233)
2. Place both palms against waist. Breathing in, slowly bend backwards. (Fig 234)
3. Try to catch ankles one after the other (Fig 235) or catch heels (Fig 236). Breathe out.
4. Maintain posture. Complete 3-5 breathing rounds.
5. Breathing out, place one arm after another against waist. (Fig 234)
6. Straighten up. (Fig 233)
7. Relax.

Half Bow posture III- *Ardha dhanurasana III*

1. Lie on tummy, arms alongside body, palms up, chin on floor. Breathe out. (Fig 237)
2. Stretch right arm forward. Breathe in. Bend right knee and catch right ankle or toes with left hand. Breathe out. (Fig 238)
3. Maintain posture. Complete 3-5 breathing rounds.
4. Lower limbs and head. (Fig 237)
5. Repeat with left arm stretched forward, right hand catching left leg.
6. Relax.

Half Bow posture IV- *Ardha dhanurasana IV*

1. Lie on tummy, arms alongside body, palms up, chin on floor. (Fig 237)
2. Stretch right arm forward. Breathe in. Bend right knee and catch right ankle or toes with left hand. Breathe out. (Fig 238)

Fig 239

Fig 240

Fig 241

Fig 242

Fig 243

Fig 244

Fig 245

Fig 246

3. Maintain posture. Complete 1-2 breathing rounds.
4. Breathing in, raise right arm and left leg a few inches off floor. Breathe out. (Fig 239)
5. Maintain posture. Complete 3-5 breathing rounds.
6. Breathing out, lower right arm and left leg. (Fig 238). Lower limbs and head. (Fig 237)
7. Repeat with left arm stretched forward, right arm catching left leg.
8. Relax.

Half Twist posture I, II, III – *Ardha matsyendra I, II, III*

1. Sit upright, legs stretched. (Fig 240)
2. Bend right knee and place right heel under left thigh. Breathe out. (Fig 241)
3. Breathe in. Cross left foot over right thigh, left ankle close to right knee. (Fig 242)
4. Maintain posture. Complete 1-2 breathing rounds.
5. Cross right arm over left knee. (Fig 243)
6. Breathe out. Catch left toes (Fig 244) **(I)** or left ankle (Fig 247) **(II)** or left knee (Fig 248) **(III)** with right hand.
7. Breathe in. Cross left arm from back to catch left knee. (Fig 245)
8. Breathing out, twist trunk and head to left. (Fig 246)
9. Maintain posture. Complete 3-5 breathing rounds.
10. Breathing in, slowly return head and trunk to front. (Fig 245)
11. Breathe out. Undo legs. (Fig 240)
12. Repeat on opposite side.
13. Relax.

Elephant posture – *Gajasana*

1. Sit on heels, toes raised, arms resting on thighs. Breathe in. (Fig 249)
2. Breathing out, bend forward, keeping head between arms. Place palms flat on floor. (Fig 250)
3. Maintain posture. Complete 1-2 breathing rounds.
4. Breathing in, raise body, flatten feet. (Fig 251)
5. Maintain posture. Complete 3-5 breathing rounds.
6. Breathing out, kneel down to Fig 250.
7. Breathe in. Sit back on heels. (Fig 249)

Fig 247

Fig 248

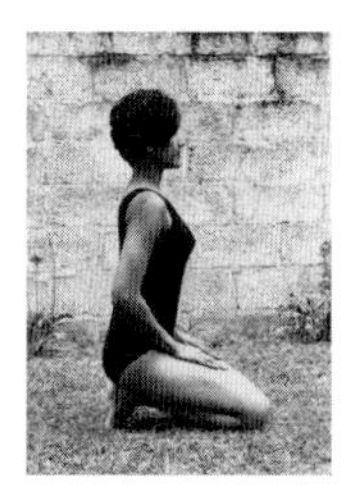

Fig 249

Fig 250

Fig 251

Fig 252

Fig 253

Fig 254

Fig 255

8. Relax.

Scalp exercises

1. Sit anywise. Hold hair in big fistfuls and gently yank forward, backward and to the sides. (Fig 252)
2. Spread fingers and press fingertips firmly on scalp. Massage scalp firmly moving fingers in circles, clockwise and anti clockwise. (Fig 253)

Eye exercise-Fixing nose tip – *Nasagra-Dhristi*

1. Sit anywise.
2. Slowly direct the eyes towards tip of nose. (Fig 254)
3. Keep gaze fixed as long as possible. Avoid strain. Breathe normally.
4. Breathing in, slowly return gaze to normal. Breathe out.
5. Close eyes and wink 5-10 times, breathing normally.
6. Relax.

Eye exercise-Fixing between eyebrows – *Bru-Madhya-Dhristi*

1. Sit anywise. Breathe out.
2. Fix tip of nose. (Fig 254)
3. Breathing in, slowly travel gaze up nose to a point between eyebrows. (Fig 255)
4. Keep fixing the point between eyebrows as long as possible. Avoid strain. Breathe normally.
5. Breathing in, slowly return gaze to normal. Breathe out.
6. Close eyes and wink 5-10 times, breathing normally.
7. Relax.

MEDITATION

CORPSE POSTURE, 5-10 minutes

Fig 256

Fig 257

Fig 258

Fig 259

Fig 260

Fig 261

Fig 262

Fig 263

ELEVENTH WEEK

Yoga is indeed amazing. Even half an hour practice daily will give a complete overhaul to our body. If we can manage more, we will certainly experience still more benefits at all levels, physical, physiological, mental and spiritual. Let us not pretext time constraints. After a while our desire to practice comes naturally and we will find ourselves doing better and better.

By now many among us may be the envy of our friends and acquaintances! Our practice is now becoming a more natural feature and part of our daily routine. Of course, we cannot afford to miss it on any account!

Fig 264

Fig 265

Fig 266

Fig 267

Fig 268

Fig 269

Fig 270

Fig 271

Legs spread forward bend – ***Prasarita Padottasana***

1. Stand legs together, hands on waist. Breathe out. (Fig 256)
2. Breathing in, spread legs wide apart, feet parallel. (Fig 257)
3. Maintain posture. Complete 1-2 breathing rounds.
4. Breathing out, bend forwards and place palms flat on floor, shoulder width apart, in line with toes. (Fig 258)
5. Breathing in, raise head, arch back. (Fig 259)
6. Maintain posture. Complete 1-2 breathing rounds.
7. Breathing out, bend to rest crown of head on floor between hands. (Fig 260, 261)
8. Maintain posture. Complete 3-5 breathing rounds.
9. Breathing in, straighten arms. (Fig 258)
10. Breathe out. Bring legs close together in one or two hops. (Fig 262)
11. Breathing in, straighten up. (Fig 256)
12. Relax.

Prayer posture – ***Pranamasana***

1. Stand legs together, arms alongside body. Breathe out. (Fig 263)
2. Breathe in. Fold arms in front of chest, palms together. (Fig 264)
3. Maintain posture. Complete 3-5 breathing rounds
4. Breathing out, lower arms to sides.(Fig 215)
5. Relax.

Palm tree posture VI – ***Talasana VI***

1. Stand legs together, arms alongside body. Breathe out. (Fig 263)
2. Breathing in, fold arms against chest. (Fig 264)
3. Maintain posture. Complete 1-2 breathing rounds.
4. Breathing in, raise arms above head, elbows straight. (Fig 265)
5. Maintain posture. Complete 3-5 breathing rounds
6. Breathing out, lower arms to sides.(Fig 263)
7. Relax.

Fig 272

Fig 273

Fig 274

Fig 275

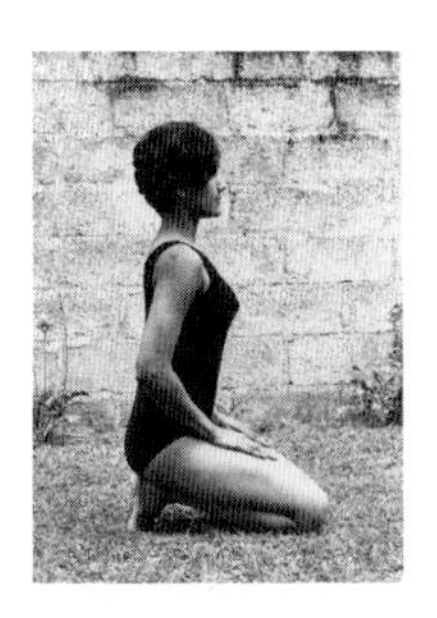

Fig 276

Fig 277

Fig 278

Fig 279

Nerve strengthening posture – *Uttanpadasana*

1. Lie on back, legs together, arms alongside body. (Fig 266)
2. Breathing out, raise head and legs off floor. Keep arms against body. (Fig 267)
3. Maintain posture. Complete 3-5 breathing rounds.
4. Breathing out, slowly lower legs, arms and head to floor. (Fig 266)
5. Relax.

Crocodile Posture III – *Makarasana III*

1. Lie down, arms spread in line with shoulders, palms down. Breathe out. (Fig 268)
2. Breathe in. Place right heel over left toes. (Fig 269)
3. Breathe out. Turn face to left.
4. Breathe in. Twist hips twice to right, both feet touching floor. (Fig 270).
5. Breathe out. Turn face to right.
6. Breathe in. Twist hips twice to left, both feet touching floor. (Fig 271)
7. Breathe out. Lower leg. (Fig 268)
8. Repeat, placing left heel over right toes.
9. Relax.

Wheel Posture – *Chakrasana*

1. Lie flat on back. Breathe out. (Fig 272)
2. Breathe in. Bend knees, heels against hips, feet flat on floor. Bend elbows and place palms on floor on either side of head. Breathe out. (Fig 273)
3. Maintain posture. Complete 3-5 breathing rounds.
4. Breathing out, push up, first legs, then trunk, (Fig 274), last head. Arch the back. (Fig 275)
5. Maintain posture. Complete 1-3 breathing rounds.
6. Breathing out, lower first head (Fig 274) then trunk, last legs. (Fig 273)
7. Lower arms and legs. (Fig 272)
8. Relax.

Fig 280

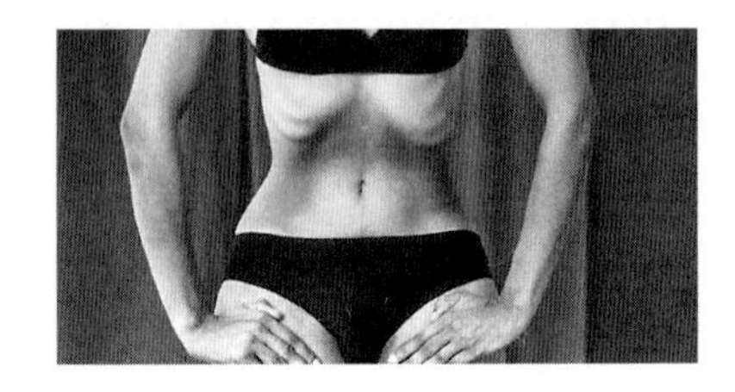

Fig 281

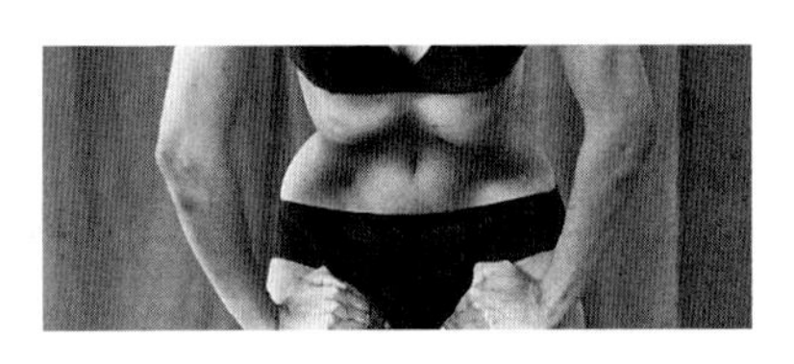

Fig 282

Fig 283

Fig 284

Fig 285

Fig 286

Fig 287

Log Posture – ***Chaturanga Dandasana***

1. Sit on heels, toes raised, arms resting on thighs. Breathe in. (Fig 276)
2. Breathing out, bend forward. Place palms flat on floor. (Fig 277)
3. Breathing in, flatten toes, raise body. (Fig 278)
4. Maintain posture. Complete 1-2 breathing rounds.
5. Breathing out, plunge forward until body is parallel to floor. (Fig 279)
6. Maintain posture. Complete 3-5 breathing rounds.
7. Breathing in, lower body. (Fig 280)
8. Breathe out. Relax.

Abdominal lift – ***Uddhiyana bandh***

1. Stand, legs shoulder width apart. Slightly arch back, place hands on thighs. Breathe out fully. (Fig 281)
2. Holding breath out, bend slightly forward. Lift in abdomen to form a hollow. (Fig 282)
3. Hold posture as long as possible, without straining.
4. Relax abdominal muscles and breathe in slowly.
5. Repeat 2-4 times.
6. Relax.

MEDITATION

CORPSE POSTURE, 5-10 minutes.

Fig 288

Fig 289

Fig 290

Fig 291

Fig 292

Fig 293

Fig 294

Fig 295

TWELFTH WEEK

For most of us it will be perhaps a first try in performing a head stand. Indeed, a thrilling experience as we advance step by step and discover how easy it is to stand on our head. To be on the safe side always practice the head stand close to a wall, about one foot away. It will be far better if we have someone to stand by in case we need help. Once mastered, we can safely practice on our own, even away from the wall.

Before attempting a head stand, rest the forehead on the floor, kneeling down to allow the blood circulation to adjust. We may use some extra padding under the head if it hurts. After a head stand, never raise head or sit up immediately. Always rest forehead down for 2 to 3 breathing rounds. Relax corpse posture for some 3-5 minutes before sitting up.

As from this week practice the abdominal lift daily. It is excellent for toning the abdominal muscles and slimming.

Tripod Headstand– *Salamba Shirshasana I*

1. Kneel down. Place hands on floor either side of knees. (Fig 283)
2. Bend head to rest on floor to form a triangle with hands. Breathe out. (Fig 284, 285)
3. Complete 3 breathing rounds.
4. Breathe in. Raise body. (Fig 286)
5. Take small steps forwards (Fig 287) until knees are level with elbows. (Fig 288)
6. Breathe out. Bring forelegs to rest in turn (Fig 289) on bent forearms. (Fig 290)
7. Maintain posture. Complete 3-5 breathing rounds.
8. Breathe out. Lower legs one by one to floor. (Fig 291)
9. Kneel down. (Fig 283). Rest forehead on hands. (Fig 292) Complete 3-5 breathing rounds. Never try to sit or stand up abruptly after head stand to avoid the risk of vertigo.
10. Relax, corpse posture, 5 minutes.

Shoulder Raising Posture – *Bahuvartenasana I*

1. Sit lotus, thunderbolt, easy posture or anywise. Rest palms on knees. Breathe out.

Fig 296

Fig 297

Fig 298

Fig 299

Fig 300

Fig 301

Fig 302

Fig 303

2. Breathe in, sharply raise and drop right shoulders, 3-5 times. (Fig 293) Breathe out.
3. Breathe in. Repeat with left shoulder. Breathe out.
4. Breathe in. Repeat with both shoulders together. (Fig 294).
5. Relax.

Shoulder Rolling Posture – *Bahuvartenasana II*

1. Sit lotus, thunderbolt, easy posture or anywise. Rest palms on knees. Breathe out.
2. Breathe in. Roll right shoulder 3-5 times, first forward (Fig 295) then roll backward 3-5 times (Fig 296). Breathe out.
3. Breathe in. Repeat with left shoulder. Breathe out.
4. Breathe in, roll both shoulders together 3-5 times, first forwards (Fig 297, 298) then backwards (Fig 299). Breathe out.
5. Relax.

Boat posture – *Navasana*

1. Lie on back, arms alongside body. Breathe out. (Fig 300)
2. Breathing in, sit up, catch ankles. (Fig 301)
3. Maintain posture. Complete 1-2 breathing rounds.
4. Breathing out straighten legs and stretch arms parallel to floor. (Fig 302)
5. Maintain posture. Complete 3-5 breathing rounds.
6. Breathing out, lie back on floor. (Fig 300)
7. Relax.

Sun salutation – *Ashtanga namaskar*

1. Sit on heels, toes raised, arms resting on thighs. (Fig 303)
2. Breathe in. Bend forward, rest palms on floor, in line with legs. (Fig 304)
3. Breathe out. Plunge to rest chin and chest on floor, between hands. (Fig 305)
4. Maintain posture for a while.
5. Breathe in. Raise chin and chest. (Fig 306)
6. Breathe out. Raise body to Fig 304.
7. Sit back. (Fig 303.) Relax.

Fig 304

Fig 305

Fig 306

Fig 307

Fig 308

Fig 309

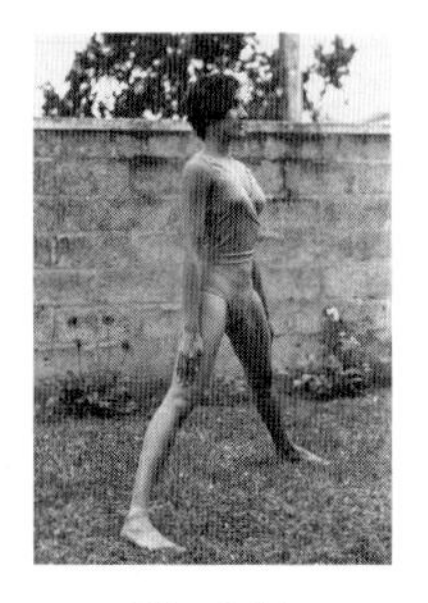
Fig 310

Fig 311

Bow Posture – *Dhanurasana*

1. Lie on tummy, arms alongside body, palms up, chin on floor. (Fig 307)
2. Bend knees and catch ankles or toes with hands. (Fig 308)
3. Maintain posture. Complete 1-2 breathing rounds.
4. Breathing in, slowly raise legs and head off floor. Keep back well arched. Breathe out. (Fig 309)
5. Maintain posture. Complete 3-5 breathing rounds.
6. Breathing out, lower legs and head (Fig 308). Straighten arms and legs (Fig 307).
7. Relax.

Half moon – *Ardha chandrasana*

1. Stand legs together, arms alongside body. Breathe out. (Fig 263)
2. Breathe in. Spread legs apart, arms alongside body. Breathe out. (Fig 310)
3. Maintain posture. Complete 1-2 breathing rounds.
4. Breathing in, raise arms in line with shoulders, palms down. Turn left leg 90 degrees to left. (Fig 311)
5. Breathing out, bend left knee to rest left palm flat on floor, level with toes. Rest right arm on body. Breathing in, turn face up. (Fig 312)
6. Complete 3-5 breathing rounds.
7. Breathing in, raise right leg. (Fig 313)
8. Maintain posture. Complete 1-3 breathing rounds.
9. Breathing out, lower right leg. (Fig 312)
10. Breathing in, straighten up to Fig 311.
11. Repeat on opposite side, turning right leg 90 degrees to right and bending right knee.
12. Relax.

MEDITATION

CORPSE POSTURE, 5-10 minutes.

Fig 312

Fig 313

Fig 314

Fig 315

Fig 316

Fig 317

Fig 318

Fig 319

THIRTEENTH WEEK

The Salutation to the sun series, Surya Namaskar, is a superb combination of fourteen postures. It may be performed independently and will amply replace a day's practice in an emergency. Over the last weeks we have already practiced all the fourteen postures separately. We will certainly have no difficulty to perform them in a single sequence. It will be an exciting and fulfilling experience.

No one should attempt Surya Namaskar before having mastered all the fourteen postures. It will be a waste of time and energy in trying the whole sequence without prior practice in stages.

Do not expect to master the sequence in one attempt. Pay particular attention to the breathing rhythm. In the beginning we may miss the exact breathing sequence. Never mind. Slowly it will come on its own.

After each round of Surya Namaskar, take a good five minutes corpse relaxation. Do deep yogic breathing.

Salutation to the Sun – *Surya Namaskar* (Stage I)

1. Stand, legs together, hands folded. (Prayer pose) (Fig 314)
2. Breathing in, raise arms above head. (Palm tree) (Fig 315)
3. Breathing out, bend forwards, place palms on floor. (Stork) (Fig 316)
4. Bend knees. (Fig 317)
5. Breathing in, stretch left leg,(Equestrian) (Fig 318) then right leg backwards. Raise body. (Elephant). (Fig 319)
6. Breathing out, plunge forward until body is parallel to floor. (Log) (Fig 320)
7. Breathing in, straighten arms at elbows and raise trunk. (Cobra) (Fig 321).
8. Breathing out, raise body up. (Elephant) (Fig 319)
9. Breathing in, bring left leg then right to rest between hands.(Squat) (Fig 322)
10. Breathing out, keeping palms on floor, raise body to (Fig 317), then to (Stork) (Fig 316).

Fig 320

Fig 321

Fig 322

Fig 323

Fig 324

Fig 325

Fig 326

Fig 327

11. Breathing in, straighten up, arms folded. (Prayer pose) (Fig 314).
12. Breathing out, lower arms alongside body. (Fig 323)
13. Relax corpse posture, 5 minutes.
14. Repeat, this time stretching right leg backwards first, then left leg.
15. Relax corpse posture, 5 minutes.

Great Seal – *Maha Mudra*

1. Sit upright, legs stretched, feet together. (Fig 323)
2. Bend right leg, to rest foot at right angle against left thigh. Lock thumbs together and place palms on left knee. (Fig 324)
3. Keep back straight. Breathing in, raise arms above head. Keep thumbs locked. (Fig 325)
4. Maintain posture. Complete 1-2 breathing rounds.
5. Breathing out, lean forward and hook left big toe with index fingers. Arch back and press chin against chest. (Chin lock) (Fig 326)
6. Maintain posture. Complete 3-5 breathing rounds.
7. Breathe in, raise arms. (Fig 325)
8. Breathing out, lower arms to knees. (Fig 324)
9. Unlock fingers. Straighten legs. (Fig 323)
10. Relax.
11. Repeat on opposite side, bending left leg.
12. Relax.

Tripod head stand II– *Salamba shirshasana II* (Stage I)

1. Kneel down. Place hands on floor either side of knees. (Fig 327)
2. Bend to rest crown of head on floor to form a triangle with hands. (Fig 328)
3. Complete 3-5 breathing rounds.
4. Breathing in, raise body. (Fig 329)
5. Complete 1-2 breathing rounds.
6. Take small steps forward (Fig 330) until trunk is perpendicular to floor (Fig 331) and legs rise off floor. (Fig 332)
7. Bend knees, toes pointing up. Breathe out. (Fig 333)

Fig 328

Fig 329

Fig 330

Fig 331

Fig 332

Fig 333

Fig 334

Fig 335

8. Complete 3-5 breathing rounds.
9. Breathing out, slowly lower legs, (Fig 332)
10. Lower feet to floor. (Fig 328)
11. Sit back on heels. (Fig 334)
12. Rest forehead on hands (Fig 334). Complete 3-5 breathing rounds. Never try to sit or stand up abruptly after head stand to avoid the risk of vertigo.
13. Relax, corpse posture, 5 minutes.

Triangular Posture – ***Trikonasana***

1. Stand legs together, arms alongside body. Breathe out. (Fig 335)
2. Breathing in, spread legs some one metre apart. Breathe out. (Fig 336)
3. Maintain posture. Complete 1-2 breathing rounds.
4. Breathing in, raise arms in line with shoulders, palms down. (Fig 337)
5. Maintain posture. Complete 1-2 breathing rounds.
6. Breathing out, bend sideways to the right until palm is flat on floor. Keep arms straight, turn face to fix left fingertips. (Fig 338).
7. For those unable to reach down, catch ankle (Fig 339).
8. Maintain posture. Complete 3-5 breathing rounds.
9. Breathing in, straighten up, keeping arms spread.(Fig 337)
10. Breathing out, lower arms. (Fig 336)
11. Repeat on opposite side.
12. Relax.

MEDITATION

CORPSE POSTURE, 5-10 minutes.

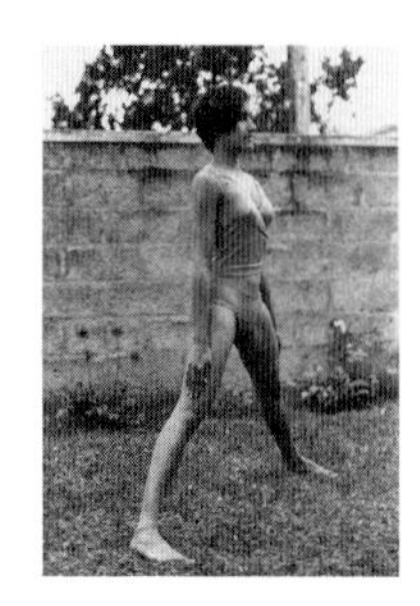

Fig 336

Fig 337

Fig 338

Fig 339

Fig 340

Fig 341

Fig 342

Fig 343

FOURTEENTH WEEK

This week we have made considerable progress in our first stage of Surya Namaskar. The breathing and posture synchronization flow more smoothly.

We will also attempt the yogic head stand. It gives a complete overhaul to the whole body. Soon we will realise the many benefits and relief these postures bring us.

Although the yogic head stand appears more difficult to perform, it will in the long run prove far easier. It is quite comfortable to support oneself upside down, in equilibrium. Suffice that we follow the instructions step by step. Remember to practice close to a wall or have someone to assist.

Fig 344

Fig 345

Fig 346

Fig 347

Fig 348

Fig 349

Fig 350

Fig 351

Repeat Surya Namaskar stage I

Yogic Headstand Posture – ***Shirshasana***

Stage I

1. Sit on raised toes. (Fig 340)
2. Interlace fingers. Bend forward, rest hands on floor. (Fig 341)
3. Place crown of head down, against interlaced fingers. (Fig 342, 343)
4. Complete 3-5 breathing rounds.
5. Breathing in, raise body. (Fig 344) Complete 1-2 breathing rounds.
6. Take small steps forward (Fig 345) until trunk is perpendicular to floor and legs rising off floor (Fig 346)
7. Bend knees, toes pointing up. (Fig 347)
8. Maintain posture. Complete 3-5 breathing rounds.
9. Breathing out, straighten knees (Fig 346) lower legs to floor. (Fig 348)
10. Sit back on heels. Rest forehead on hands. (Fig 349) Complete 3-5 breathing rounds. Remember, we should never try to sit or stand up abruptly after head stand to avoid the risk of vertigo.
11. Relax, corpse posture, 5 minutes.

MEDITATION

CORPSE POSTURE, 5-10 minutes.

Fig 352

Fig 353

Fig 354

Fig 355

Fig 356

Fig 357

Fig 358

Fig 359

FIFTEENTH WEEK

This week will be devoted to the mastery of the Surya Namaskar. It will be quite hectic at the outset, especially to synchronise breathing with the postures. But it will be worth the effort. It is an amazing combination of postures for perfect health.

Surya namaskar stimulates almost every system in our body – the cardiovascular system because it makes the heart stronger, the digestive system as well as the nervous system. It also improves the functioning of the endocrine glands like the thyroid, parathyroid and pituitary glands.

Head stand comes as a blessing from the blue. There is reduced pressure on the veins. The shoulder stand, head stand and inverse postures should be regularly practiced, if not daily. In these postures the blood flows back to the heart with less effort. It will be a great relief to the blood vessels and their owner! These postures give the most astounding results. The fragile blood vessels are rejuvenated. In many cases varicose veins may be healed.

Salutation to Sun – *Surya Namaskar* Complete

1. Stand, legs close together, hands folded. (Prayer pose) (Fig 350)
2. Breathing in, raise arms above head. Lock fingers, turn palms up. (Palm tree) (Fig 351). Bend backwards.(Crescent) (Fig 352)
3. Breathing out, bend forwards, place palms on floor. (Stork) (Fig 353)
4. Breathing in, send right leg behind.(Equestrian)(Fig 354)
5. Breathing out, raise head. (Equestrian) (Fig 355)
6. Breathing in, raise arms up. (Crescent lunge) (Fig 356)
7. Breathing out, lower arms, palms on floor. (Equestrian) (Fig 354)
8. Breathing in, send left leg back, raise body. (Elephant) (Fig 357)
9. Breathing out, plunge forward, body parallel to floor, (Log) (Fig 358).
10. Breathing in, raise trunk (Cobra). (Fig 359)

Fig 360

Fig 361

Fig 362

Fig 363

Fig 364

Fig 365

Fig 366

Fig 367

11. Breathing out, raise body (Elephant). (Fig 357)
12. Breathe in. Kneel down (Cat) (Fig 360).
13. Breathing out, rest chin and chest on floor. (Ashtanga) (Fig 361)
14. Breathing in, raise trunk (Cobra). (Fig 362)
15. Breathing out, raise body. (Elephant) (Fig 357)
16. Breathing in, send right leg forward. (Equestrian) (Fig 363)
17. Breathing out, raise head. (Equestrian) (Fig 364)
18. Breathing in, raise arms up. (Crescent lunge) (Fig 365).
19. Breathing out, lower arms, palms on floor. (Equestrian) (Fig 363)
20. Breathing in, send right leg back. (Elephant) (Fig 357)
21. Breathing out, plunge forward, body parallel to floor. (Log) (Fig 358).
22. Breathing in, raise trunk (Cobra). (Fig 359)
23. Breathing out, raise body. (Elephant) (Fig 357)
24. Breathing in, kneel down (Cat) (Fig 365).
25. Breathing out, rest chin and chest on floor. (Ashtanga) (Fig 361)
26. Breathing in, raise trunk (Cobra). (Fig 362)
27. Breathing out, raise body. (Elephant) (Fig 357)
28. Breathe in, kneel down. (Cat) (Fig 360)
29. Breathe out, sit on heels (Thunderbolt)(Fig 366)
30. Breathe in, raise arms. (Thunderbolt) (Fig 367)
31. Breathing out, bend forwards, stretch arms on floor. (Swan) (Fig 368)
32. Breathing in, raise body. (Elephant) (Fig 357)
33. Breathing out, plunge forward, body parallel to floor. (Log) (Fig 358).
34. Breathing in, raise trunk (Cobra) (Fig 359).
35. Breathing out, raise body. (Elephant) (Fig 357)
36. Breathing in. Kneel down (Cat) (Fig 360).
37. Breathing out, rest chin and chest on floor (Ashtanga) (Fig 361)
38. Breathing in, raise trunk (Cobra). (Fig 362)
39. Breathing out, raise body. (Elephant) (Fig 357)
40. Breathing in, first send right leg, then left leg in front to squat.(Squat) (Fig 369)
41. Breathing out, straighten knees, raise body halfway. (Stork) (Fig 353)
42. Breathing in, straighten up, hands folded. (Prayer pose) (Fig 350).
    This completes one of the two sets that make up a complete Surya Namaskar.
43. Relax, Corpse at least 5 minutes.

Fig 368

Fig 369

Fig 370

Fig 371

Fig 372

Fig 373

Fig 374

Fig 375

44. Complete the second set, this time sending left leg behind first.

For those who cannot cope with two sets at a time there is no harm to perform one set one day and the other set, starting with the other leg the following day.

Tripod head stand II- *Salamba Shirshasana II*
Stage 2

1. Kneel down. Place hands on floor either side of knees. (Fig 370)
2. Bend to rest crown of head on floor to form a triangle with hands. (Fig 371)
3. Complete 3-5 breathing rounds.
4. Breathing in, raise body. (Fig 372) Take small steps forward (Fig 373) until trunk is perpendicular to floor (Fig 374) and legs rise off floor. (Fig 375)
5. Bend knees, toes pointing up. Breathe out. (Fig 376)
6. Complete 1-2 breathing rounds.
7. Raise thighs parallel to floor. (Fig 377)
8. Maintain posture. Complete 3-5 breathing rounds.
9. Breathing out, keeping knees bent, slowly lower thighs to Fig 376.
10. Breathing out, straighten knees. (Fig 375) Lower legs to floor.(Fig 371)
11. Sit back on heels. Rest forehead on hands. (Fig 378) Complete 3-5 breathing rounds. Remember, we should never try to sit or stand up abruptly after head stand to avoid the risk of vertigo.
12. Relax, corpse posture, 5 minutes.

MEDITATION

CORPSE POSTURE 5-10 minutes

Fig 376

Fig 377

Fig 378

Fig 379

Fig 380

Fig 381

Fig 382

Fig 383

SIXTEENTH WEEK

We are nearing the completion of our course. As we have discovered, there is nothing strenuous about the postures. Easy to perform and easy to remember! Yoga gives us a great stretch and keeps us fit physically. It is extremely beneficial for our joints, ligaments and improves flexibility and enhances carriage. It also does wonders for our mental and emotional health.

We should keep on practicing the Surya Namaskar daily. The earlier it is mastered, the better. The real pleasure will be to perform the complete set from memory. It will save us precious time. In the beginning we should concentrate on one set daily. We should alternate sets, one day starting on the right leg, the next day starting on the left. Soon we will be able to complete both sets to make a complete Surya Namaskar.

Repeat one set Surya Namaskar.

Tripod head stand II – *Salamba Shirshasana II*
Complete

1. Kneel down. Place hands on floor either side of knees. (Fig 370)
2. Bend to rest crown of head on floor to form a triangle with hands. (Fig 371)
3. Complete 3-5 breathing rounds.
4. Breathing in, raise body. (Fig 372) Take small steps forward (Fig 373) until trunk is perpendicular to floor (Fig 374) and legs rise off floor. (Fig 375)
5. Bend knees, toes pointing up. Breathe out. (Fig 376)
6. Complete 1-2 breathing rounds.
7. Raise thighs parallel to floor. (Fig 377)
8. Raise thighs in line with body. Keep knees bent. (Fig 379)
9. Maintain posture for 1-2 breathing rounds.
10. Breathing in, slowly straighten legs, toes pointing up. (Fig 380)
11. Maintain posture for 3-5 breathing rounds.
12. Breathing out, reverse steps to come down, first bending knees to Fig 379, then to Fig 377 then to Fig 376. Slowly lower legs to Fig 375 to rest knees on floor (Fig 371).
13. Sit back on heels. Rest forehead on arms. (Fig 378). Complete 3-5 breathing rounds. Never try to sit or stand up abruptly after head stand to avoid the risk of vertigo.
14. Relax corpse posture, at least 5 minutes.

MEDITATION

Corpse posture, 5-10 minutes.

SEVENTEENTH WEEK

We are enjoying our progress in the head stand postures. Some may need more practice, but we must persevere. We will succeed. Do not stay more time than indicated; 30 seconds are amply sufficient for a beginner. Never go beyond 2-3 minutes in head stand, even when fully proficient.

Yoga is not a chore to be completed. One should undertake it with all joy. We should do our part and not feel disappointed if our performance is wanting. Persevere! Only then will we experience the full benefits of yoga. Our body will be more flexible, lighter, younger; our mind calm and composed.

Repeat one set Surya Namaskar.

Fig 384

Fig 385

Fig 386

Fig 387

Fig 388

Fig 389

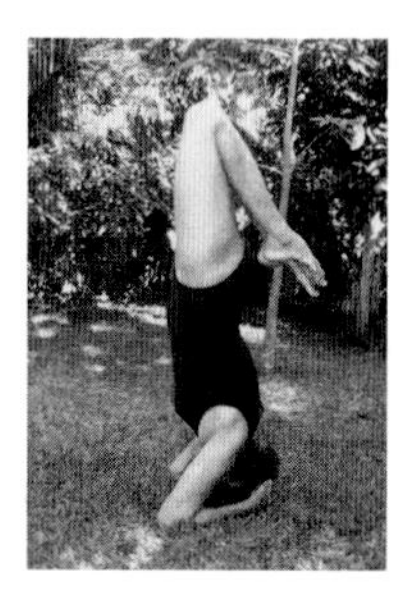

Fig 390

Fig 391

Yogic Head stand – *Shirshasana*
Stage II

1. Sit on heels. Breathe in. (Fig 381)
2. Interlace fingers. Bend forward, rest hands on floor. (Fig 382)
3. Place crown of head down, against interlaced fingers. Breathe out.(Fig 383)
4. Complete 3-5 breathing rounds.
5. Breathing in, raise body. (Fig 384)
6. Take small steps forward (Fig 385) until trunk is perpendicular to floor and legs rising off floor. (Fig 386)
7. Bend knees, toes pointing up. (Fig 387) Complete 1-2 breathing rounds.
8. Slowly raise thighs until parallel to floor. (Fig 388)
9. Maintain posture for 3-5 breathing rounds.
10. Breathing out, reverse steps to come down, first bending knees to Fig 387. Slowly lower legs to Fig 386 to rest knees on floor (Fig 383).
11. Sit back on heels. Rest forehead on arms. (Fig 389). Complete 3-5 breathing rounds. Never try to sit or stand up abruptly after head stand to avoid the risk of vertigo.
12. Relax corpse posture, at least 5 minutes.

MEDITATION

Corpse posture 5-10 minutes.

EIGHTEENTH WEEK

Congratulations! We are completing the last of our 108 postures, in all satisfaction.

As from next week, we will be embarking on our permanent daily series. We have to go through them with the same attention until perfectly mastered. For postures where performance has not been satisfactory, we will have to carry on until we are perfect. Anyway, yoga practice is for a life time. Those who can manage easily may now perform the two sets of a complete Surya namaskar.

If performed according to rules, yoga does not strain or cause injury. A morning session relieves stiffness, revitalises our body and refreshes our mind. Done during the day, it will instantly boost us. Performed after sunset it will help us unwind. Each day's practice will bring its own pleasure, satisfaction and an inner joy.

GOOD LUCK!

Repeat Surya Namaskar.

Yogic Head stand – *Shirshasana*
Complete

1. Sit on heels. Breathe in. (Fig 381)
2. Interlace fingers. Bend forward, rest hands on floor.(Fig 382)
3. Place crown of head down against interlaced fingers. Breathe out.(Fig 383)
4. Complete 3-5 breathing rounds.
5. Breathing in, raise body. (Fig 384)
6. Take small steps forward (Fig 385) until trunk is perpendicular to floor and legs rising off floor. (Fig 386)
7. Bend knees, toes pointing up. (Fig 387) Complete 1-2 breathing rounds.
8. Slowly raise thighs until parallel to floor. (Fig 388)
9. Maintain posture for 1-2 breathing rounds.
10. Raise thighs in line with body. Keep knees bent. (Fig 390)
11. Breathing in, slowly straighten legs, toes pointing up. (Fig 391)
12. Maintain posture for 3-5 breathing rounds.
13. Breathing out, reverse steps to come down, first bending knees to Fig 390, then to Fig 388 then to Fig 387. Slowly lower legs to Fig 386 to rest knees on floor (Fig 383).
14. Sit back on heels. Rest forehead on arms. (Fig 389). Complete 3-5 breathing rounds. Never try to sit or stand up abruptly after head stand to avoid the risk of vertigo.
15. Relax corpse posture, at least 5 minutes.

MEDITATION

Corpse posture, 5-10 minutes.

DAILY YOGA SERIES

As from this week we are starting with the permanent daily series. From Monday to Saturday, different series have been worked out. Postures like warm ups, lion posture, abdominal lift etc, are to be practiced daily.

The daily series have been designed to spread over four weeks. Once the four weeks' sequence completed, we start over again. When we are used to the daily series we can if desired work out new schedules.

Surya namaskar is in itself a complete series and can be practiced independently. The postures present in the surya namaskar have not been included in the daily series. For this reason the surya namaskar has to be practiced regularly. Relaxation postures too, are a complete set by themselves. They can replace a normal day's practice. They have not been included in the daily series since they replace corpse posture for relaxation between daily postures, sitting, standing or lying down. Many relaxation postures also serve as basis for other postures.

The breathing and postures involve body and mind. The mental repetition will enhance concentration and prevents the mind from being distracted. At the end of the daily series follows a ten minute meditation. We should never miss this portion of the daily practice, as within this lies the consolidation of our daily practice. The subtle life force, prana will become more and more manifest to enhance our personal efforts.

DAILY

Night

1. Scalp exercises
2. Leg stretching, (also on waking up)

Anytime

1. Neck roll
2. Eye exercises
3. Face exercises.

MONDAY I

1. Palm tree I
2. Palm tree II
3. Stick
4. Lion I
5. Finger tension
6. Surya Namaskar
7. Yogic Head stand
8. Shoulder stand
9. Shoulder rolling
10. Abdominal lift
11. Meditation
12. Corpse

TUESDAY I

1. Palm tree III
2. Palm tree IV
3. Stick
4. Lion II
5. Finger tension
6. Reverse
7. Sideways stretch
8. Locust I
9. Knee bend I
10. Abdominal lift
11. Meditation
12. Corpse

WEDNESDAY I

1. Palm tree I
2. Palm tree II
3. Stick
4. Lion I
5. Finger tension
6. Surya Namaskar
7. Tripod Head stand
8. Shoulder stand
9. Shoulder raising
10. Abdominal lift
11. Meditation
12. Corpse

THURSDAY I

1. Palm tree III
2. Stick
3. Lion II
4. Finger tension
5. Half locust
6. Half bow I
7. Fish I
8. Crocodile II
9. Leg spread forward bend
10. Abdominal lift
11. Meditation
12. Corpse

FRIDAY I

1. Palm tree I
2. Palm tree II
3. Stick
4. Lion I
5. Finger tension
6. Surya Namaskar
7. Yogic Head stand
8. Shoulder stand
9. Shoulder rolling
10. Abdominal lift
11. Meditation
12. Corpse

SATURDAY I

1. Palm tree III
2. Palm tree IV
3. Stick
4. Lion II
5. Finger tension
6. Catching big toe
7. Head to knee
8. Half bow II
9. Angle balance
10. Abdominal lift
11. Meditation
12. Corpse

MONDAY II

1. Palm tree I
2. Palm tree II
3. Stick
4. Lion posture I
5. Finger tension
6. Fish II
7. Locust II
8. Cat
9. Tree
10. Abdominal lift
11. Meditation
12. Corpse

TUESDAY II

1. Palm tree III
2. Palm tree IV
3. Stick
4. Lion II
5. Finger tension
6. Surya Namaskar
7. Tripod Head stand
8. Shoulder stand
9. Shoulder raising
10. Abdominal lift
11. Meditation
12. Corpse

WEDNESDAY II

1. Palm tree I
2. Palm tree II
3. Stick
4. Lion I
5. Finger tension
6. Twist
7. Catching both toes
8. Forward bend
9. Wedge
10. Abdominal lift
11. Meditation
12. Corpse

THURSDAY II

1. Palm tree III
2. Palm tree IV
3. Stick
4. Lion II
5. Finger tension
6. Surya Namaskar
7. Yogic Head stand
8. Shoulder stand
9. Shoulder rolling
10. Abdominal lift
11. Meditation
12. Corpse

FRIDAY II

1. Palm tree I
2. Palm tree II
3. Stick
4. Lion I
5. Finger tension
6. Child
7. Twisting cross
8. Half cobra I
9. Triangular
10. Abdominal lift
11. Meditation
12. Corpse

SATURDAY II

1. Palm tree III
2. Palm tree IV
3. Stick
4. Lion II
5. Finger tension
6. Surya Namaskar
7. Tripod Head stand
8. Shoulder stand
9. Shoulder raising
10. Abdominal lift
11. Meditation
12. Corpse

MONDAY III

1. Palm tree I
2. Palm tree II
3. Stick
4. Lion I
5. Finger tension
6. Fish III
7. Locust III
8. Cow
9. Palm tree VI
10. Abdominal lift
11. Meditation
12. Corpse

TUESDAY III

1. Palm tree III
2. Palm tree IV
3. Stick
4. Lion II
5. Finger tension
6. Surya Namaskar
7. Yogic Head stand
8. Shoulder stand
9. Shoulder rolling
10. Abdominal lift
11. Meditation
12. Corpse

WEDNESDAY III

1. Palm tree I
2. Palm tree II
3. Stick
4. Lion I
5. Finger tension
6. Half twist
7. Plough
8. Squat
9. Nerve strengthening
10. Abdominal lift
11. Meditation
12. Corpse

THURSDAY III

1. Palm tree III
2. Palm tree IV
3. Stick
4. Lion II
5. Finger tension
6. Surya Namaskar
7. Tripod Head stand
8. Shoulder stand
9. Shoulder rolling
10. Abdominal lift
11. Meditation
12. Corpse

FRIDAY III

1. Palm tree I
2. Palm tree II
3. Stick
4. Lion I
5. Finger tension
6. Tortoise
7. Fish shake
8. Half cobra
9. Half moon
10. Abdominal lift
11. Meditation
12. Corpse

SATURDAY III

1. Palm tree III
2. Palm tree IV
3. Stick
4. Lion II
5. Finger tension
6. Surya Namaskar
7. Yogic Head stand
8. Shoulder stand
9. Shoulder raising
10. Abdominal lift
11. Meditation
12. Corpse

MONDAY IV

1. Palm tree I
2. Palm tree II
3. Stick
4. Lion I
5. Finger tension
6. Surya Namaskar
7. Tripod Head stand
8. Shoulder stand
9. Shoulder rolling
10. Abdominal lift
11. Meditation
12. Corpse

TUESDAY IV

1. Palm tree III
2. Palm tree IV
3. Stick
4. Lion II
5. Finger tension
6. Wheel
7. Sideways stretch II
8. Camel
9. Knee bend II
10. Abdominal lift
11. Meditation
12. Corpse

WEDNESDAY IV

1. Palm tree I
2. Palm tree II
3. Stick
4. Lion I
5. Finger tension
6. Surya Namaskar
7. Yogic Head stand
8. Shoulder stand
9. Shoulder raising
10. Abdominal lift
11. Meditation
12. Corpse

THURSDAY IV

1. Palm tree IV
2. Stick
3. Lion II
4. Finger tension
5. Locust all up
6. Half bow III
7. Bow
8. Crocodile III
9. Lateral limbs spread
10. Abdominal lift
11. Meditation
12. Corpse

FRIDAY IV

1. Palm tree I
2. Palm tree II
3. Stick
4. Lion I
5. Finger tension
6. Surya Namaskar
7. Tripod Head stand
8. Shoulder stand
9. Shoulder rolling
10. Abdominal lift
11. Meditation
12. Corpse

SATURDAY IV

1. Palm tree III
2. Palm tree IV
3. Stick
4. Lion II
5. Finger tension
6. Pump
7. Great seal
8. Half bow IV
9. Boat
10. Crocodile
11. Abdominal lift
12. Meditation
13. Corpse

THE MARVEL OF MARVELS

We have reached the end of our course. There should be many smiles of satisfaction and pride too! So many, for too long may have underestimated their talent or capability. Today we are transformed, rejuvenated, and full of vitality. We have recovered our youth!

After these few weeks of practice we would have gained in poise and confidence. We know now what real exercise means. Not fatigue and discomfort or heavy sweating! In yoga the movements are graceful, without any strain and the end result is simply revitalization, reinvigoration, not exhaustion. All components of our being no longer fragmented, now work in unison. Our long lost unity reintegrated. Mind, body and spirit reunited. Harmony, health and happiness rediscovered.

Very often we mistake our daily frantic activity whether at home or at work as equivalent to 'physical exercise'! We forget these daily repetitive actions are at the very basis of all our physical, mental and emotional stress. An occasional visit to a spa or massage parlour, or some sports on odd days or 'relaxation' in front of the TV, will certainly not remedy our dire situation.

Our daily yoga practice is the ideal choice. Learning to work the whole body and discovering complete yogic relaxation will relieve us of all tension. It will revitalize us at no cost, in the comfort of our home sweet home! No longer are we bothered by the minor ills caused by work pressure or environmental fluctuations. Colds, flu and allergies are things of the past.

By our progress during these weeks of sincere application we will be the best judge of the marvel of yoga. Whatever our age, our physical condition, yoga and meditation remain the ideal for integral balance: PERFECT HOMEOSTASIS.

INDEX

Posture	**Sanskrit name**	**Week**	**Page**
Crescent	*Ardha Chandrasana*	4	103
Crescent lunge I	*Anjanaya asana I*	5	109
Crescent lunge II	*Anjanaya asana II*	5	111
Crocodile I	*Makarasana I*	1	83
Crocodile II	*Makarasana II*	4	103
Crocodile III	*Makarasana III*	11	157
Deep diaphragmatic breathing in Corpse	*Deergha Swasam*	1	81
Deep diaphragmatic breathing - Sitting	*Deergha Swasam*	1	85
Easy Posture	*Sukhasana*	1	83
Elephant	*Gajasana*	10	149
Equestrian I	*Ashwa sanchalanasana*	5	109
Equestrian II	*Ashwa sanchalanasana II*	5	109
Eye exercises	*Netra Vyayamam*	9	143
Eye exercises – Fixing nose tip	*Nasagra-Dhristi*	10	151
Eye exercises – Fixing between eyebrows	*Bru-Madhya-Dhristi*	10	151
Face exercises		7	129
Finger tension	*Hastapawanmukta*	6	121
Fish I	*Matsyasana I*	6	119
Fish II	*Matsyasana II*	8	131
Fish III	*Matsyasana III*	9	139

Posture	**Sanskrit name**	**Week**	**Page**
Wheel	*Chakrasana*	11	157
Yogic breathing		3	97

SELECTED POSTURES FOR COMING BOOKS

THE KEY TO HAPPINESS IS IN OUR HANDS. WE SHOULD NOT MISLAY IT.

ALL THE WEALTH OF THE WORLD WILL NOT BUY US AN IOTA OF HEALTH, HARMONY OR HAPPINESS.

HEALTH MAY NOT BRING US MUCH WEALTH, BUT PLENTY OF HAPPINESS AND HARMONY.

THIS IS THE MIRACLE OF YOGA!